Rural Granaries in Northern Gaul (6th Century BCE–4th Century CE)

Radboud Studies in Humanities

VOLUME 8

The titles published in this series are listed at *brill.com/rsh*

Rural Granaries in Northern Gaul (6th Century BCE–4th Century CE)

From Archaeology to Economic History

Edited by

Stéphane Martin

BRILL

LEIDEN | BOSTON

Cover illustration: Reconstruction by H. Mylius of building VI, interpreted as a granary, on the Köln-Müngersdorf villa (extracted from F. Fremersdorf, *Der römische Gutshof Köln-Müngersdorf*, Berlin, 1933, Taf. 13).

The Library of Congress Cataloging-in-Publication Data is available online at http://catalog.loc.gov
LC record available at http://lccn.loc.gov/

Typeface for the Latin, Greek, and Cyrillic scripts: "Brill". See and download: brill.com/brill-typeface.

ISSN 2213-9729
ISBN 978-90-04-38903-8 (hardback)
ISBN 978-90-04-38904-5 (e-book)

This book is printed on acid-free paper and produced in a sustainable manner.

Contents

Figures and Tables

Figures

Tables

Notes on Contributors

Stanislas Bossard
MA (2015), University of Nantes, is currently a PhD student at this university. His research focuses on the Celtic and Roman period in Western France and he works in particular on sanctuaries and storage in Gallic farmsteads. His Master thesis (*Les souterrains gaulois en Bretagne et Normandie occidentale*) will be published soon in the "Archéologie & Culture" series of the Presses universitaires de Rennes.

Lars Blöck
PhD (2012), is currently working at the Rheinisches Landesmuseum Trier. His doctoral dissertation, dealing with the settlement history of a region corresponding roughly to south-western Baden-Württemberg, as recently been published: *Die römerzeitliche Besiedlung im rechten südlichen Oberrheingebiet* (Wiesbaden, 2016)

Alain Ferdière
Emeritus Professor of Archaeology, Université de Tours, is a recognized expert of the rural history and archaeology of Roman Gaul. Among his publications are *Les campagnes en Gaule romaine* (*52 av. J.-C.-486 ap. J.-C.*) (Paris, 1988, 2 vols.) and *Les Gaules, Provinces des Gaules et Germanies, Provinces Alpines. IIe siècle av.-Ve siècle ap. J.-C.* (Paris, 2005).

Stéphane Martin
PhD (2013), was postdoctoral researcher in the frame of the Anchoring Innovation project at Radboud Universiteit Nijmegen from 2015 to 2018, before working for the ERC project RurLand in 2018. His doctoral dissertation, published in 2015, was concerned with the Romanization and monetization of northern Gaul: *Du statère au sesterce. Monnaie et romanisation dans la Gaule du Nord et de l'Est* (*IIIe s. a.C.–Ier s. p.C.*) (Bordeaux, 2015). He recently edited a collective volume about money and coins in the Gallic countryside: *Monnaies et monétarisation dans les campagnes de la Gaule du Nord et de l'Est, de l'Âge du Fer à l'Antiquité tardive* (Bordeaux, 2016).

Michel Reddé
Directeur d'études, École Pratique des Hautes Études, Paris, was the principal investigator of the project RurLand (Rural Landscapes in Roman north-eastern Gaul), funded by the European Research Council (2014-2018). The results of the

project have been recently published: Gallia Rustica. *Les campagnes du nord-est de la Gaule, de la fin de l'âge du Fer à l'Antiquité tardive* (Bordeaux, 2 vols., 2017 and 2018). Michel Reddé is also an internationally recognized expert of Roman military history and archaeology.

Javier Salido Domínguez
PhD (2010), is currently Researcher "Juan de la Cierva-Incorporación"- Ministry of Economy and Competitiveness of Spain, at Universidad Complutense de Madrid. He has worked extensively on grain production and storage in the western part of the Roman world, publishing several books on the subject, including Horrea militaria. *El aprovisionamiento de grano al ejército en el occidente del Imperio Romano* (Madrid, 2011) and *Arquitectura rural romana: graneros y almacenes en el Occidente del Imperio* (Autun, 2017).

Introduction

Stéphane Martin

In a paper published in 2012 but written in 2006, François Sigaut reviewed the progress made in the study of food storage since the three pioneering volumes he had edited with Marceau Gast and other colleagues (Sigaut 2012; Gast and Sigaut 1979; Gast, Sigaut, and Bruneton-Governatori 1981; Gast et al. 1985).[1] He noted that archaeologists had been eager to tackle the problem of storage, anthropologists less so, and historians not at all. Fortunately, he was being proved wrong at the very same time as he was writing those lines (see below for a review of the literature). Food storage is a hot topic again in current research, and for once scholars of the ancient world are at the forefront.

There are good reasons for this. It is well known that scale and intensity of the transport of staples are good indicators of the degree of economic integration of a given region within the ancient world. Of all foodstuffs, grain was probably the most important in ancient times, both in terms of volume and because it was so central (and still is in most countries) to the various diets in and around the Mediterranean area. Now storage intervenes at all stages in the production and distribution of grain, as made clear e.g. by Enim Tengström's study of late antique *annona* (Tengström 1974). It is therefore no exaggeration to write that "*la question des structures de stockage est au cœur des débats des spécialistes sur la nature de l'économie antique*" (ANR Entrepôts 2012).

1 Previous Research on Grain Storage in Antiquity: a Bias towards Mediterranean Towns

In the last quarter of the twentieth century, Geoffrey Rickman's monographs remained the only work on the technical and practical aspects of grain storage under the Roman Empire (Rickman 1971, 1980. During the same period, institutional and legal aspects received more attention: see above all Sirks 1991, with a review of previous literature). No equivalent existed for the Greek world. But the early twenty-first century witnessed an upsurge in publications, above

1 Since they are generally ignored outside of the French-speaking world (in spite of at least one very important paper published in English: Sigaut 1988), it is worth reminding the reader that Sigaut's scientific output is freely available online: <http://www.francois-sigaut.com> (accessed 7 June 2018).

all from French and Italian archaeologists and historians. This was the result of a number of French and European research programmes led by Catherine Virlouvet and Brigitte Marin, most notably the RAMSES2 project "Le ravitaillement des villes du bassin méditerranéen de l'Antiquité aux temps modernes" from 2006 to 2008, followed from 2009 to 2012 by the ANR Project "Entrepôts et lieux de stockage du monde gréco-romain antique" (Marin and Virlouvet 2004, 2008a, 2008d, 2008b, 2008c, 2016; on Africa alone: Virlouvet 2007; with a focus on Spain: Arce and Goffaux 2011. See most recently Chankowski, Lafon, and Virlouvet 2018. Stemming from another project, Erdkamp 2005, 143–68, has a focus on the economic significance of storage rather than the technical and practical sides. See also Kloft 2016). Thanks to these projects, various teams of archaeologists and historians have worked on storage and distribution of agricultural goods (mainly grain) of various pre-industrial economies in the *longue durée*, from the Classical world (with a focus on ancient Rome) to the end of the eighteenth century. The main problem was clearly defined: to gain insight into the logistics of the food supply of non-rural communities, mainly towns, and above all Rome in ancient and modern times. From the outset, this specific research theme has very much oriented new research on storage. It is striking to note that the numerous publications almost exclusively adopt the point of view of urban (and to a lesser extent military) consumers, not of rural producers. The problem of storage in the grain-producing regions is barely addressed; when it is, the focus is on the regions supplying Rome, i.e., for the Principate, Africa and Egypt. Furthermore, although the study area of the ANR "Entrepôts" project encompassed the whole of the Roman Empire, there is actually a strong focus on the Mediterranean.

While much of what has been written is specific to urban storage, the following points can also apply to storage in a rural context. In particular, it is worth mentioning a paper by Virlouvet in which she attempts to draw up a typology of storage buildings based not on their architecture, but on their function in the distribution and supply chain (Virlouvet 2011; expanded in Bernardos Sanz and Virlouvet 2016). She stresses rightly that an architectural type cannot be attributed with certainty to a specific function. The point had already been made by Sigaut (1988, 10) regarding granaries on posts, noting that similar looking structures may have very different functions – the converse is of course also true. This is all the more important since archaeologists generally excavate only the foundations of storage buildings and, in the case of post-elevated structures, even shallower traces (Groenewoudt 2011; for orchards mistaken as granaries: Koehler 2003; for post-built hypocausts, also looking like granaries: Lehar 2015). In emphasising function, Virlouvet also reminds us that the role of the storage structure in the chain between producer and consumer is crucial to its understanding. The nature of the stored produce and the duration of storage directly influence the design of the building and how we should study it and calculate its capacities.

Furthermore, written sources make it very clear that there is no connection between the size of a storage building and the identity of its owner (Virlouvet 2011, 12–14). Contrary to what is often written, small does not mean private and big is not synonymous with public, even in towns. What's more, the owner of the building need not be the owner of the produce stored in it: a private individual could store produce in public *horrea* and public grain could be stored in private granaries (Durliat 1990, 463–71; Dubouloz 2008. For private granaries rented by public authorities in modern Lyon, see Martinat 2008; Nicolier 2015).

If finally we turn to storage in the countryside, a recent review by J. Salido Domínguez (2008; 2017, 18–23) makes it clear that the existing literature stems almost exclusively from archaeologists (for the state-of-the-art on Iron Age storage in Gaul, see Chapter 4 by Bossard in this volume). The problem was first tackled through the identification of storage buildings, in particular granaries, with a focus on the north-western provinces (for Britain: Morris 1979, 29–39, with Black 1981. For the Gauls and Germanies, see e.g.: Demarez 1987, 5–10; Van Ossel 1992, 154–59; Schinkel 1998, 255–66; most recently, Ferdière 2015; Schubert 2016). In recent years, archaeobotanical data have received increased attention and are often combined with architectural typology to identify storage sites (to cite but a few names, see works by Marijke van der Veen for Britain, Laura I. Kooistra for the Netherlands, Véronique Zech-Matterne for France).

Interest in rural storage dates back to the 1970s, just as for urban and military storage. But contrary to the latter two, it has not managed to raise much interest among historians and is still missing from recent work on the ancient economy. See e.g. the recent *The Roman Agricultural Economy. Organisation, Investment, and Production* edited by Alan Bowman and Andrew Wilson (2013b). Through a series of case-studies, the book works towards a bottom-up quantitative approach grounded in actual data rather than the usual top-down modelling approach. Although it makes no claim to exhaustivity, it is still surprising not to find any study of storage buildings and no attempt at quantifying their capacity. Bowman and Wilson (2013a, 3) are surely right when they explain the shortage of studies on grain compared to wine and olive oil, by the fact that "milling and storage facilities for cereals have proved harder to exploit" – although one wonders if the lack of interest in rural granaries in the Mediterranean area is not an equally important reason (for Italy, Rossiter 1978, the only attempt at a general overview on farm buildings, devotes only pp. 57–59 to granaries; now supplemented by Busana 2002, 193–203 for Veneto, and Pellegrino 2017 for the Vesuvian area. For Spain, see Salido Domínguez 2017; but the number of granaries remains very low compared to Gaul studied in the same monograph).

2 From Grain Storage to Grain Production: the Case of Northern Gaul

For all the difficulties we may encounter, it remains absolutely crucial to engage with grain storage facilities found in the countryside. As acknowledged by the directors of the ANR Project "Entrepôts et lieux de stockage du monde gréco-romain antique" themselves, a better appraisal of rural storage is needed to move forward on the subject (Chankowski, Lafon and Virlouvet 2018, 272–73).

Sigaut has emphasised on several occasions that storage is central not only to grain distribution, but also to grain production. A storage facility will be built according to how the grain is harvested, how and when it is threshed, in what form and at what stage it is stored, and so on. Therefore, studying storage is much more than understanding how grain was preserved; it means, above all, trying to understand the nature of the agrarian system that needed these storage buildings.

Furthermore, for all its pitfalls, estimating storage capacities is about the only way we have of attempting a quantification of cereal production from ancient data. Grain itself has only been preserved in minute quantities; pollen analysis can help estimate the intensity of grain production but will not yield any quantitative insights; outside Egypt, textual sources offer no concrete data on the subject[2]. Storage buildings remain the only hard evidence at hand. What's more, they are virtually present in every part of the Roman Empire and as such could be the starting-point of an inter-provincial study. However, an Empire-wide study is inconceivable at the moment: on the one hand, the quantity of available data is overwhelming; on the other, lack of interest in rural storage facilities makes a systematic comparison of several, if not all, provinces impossible. These considerations dictated the choice of northern Gaul[3] as our study area.

2 Let us note in passing that the wealth of textual evidence from Egypt is not matched in the archaeological record : well-identified and studied granaries appear to be very rare according to the recent survey in Geraci 2008 (on both archaeological structures and texts).

3 In this volume, northern Gaul will often be used as a shorthand for the northern part of the *Tres Galliae* (*Gallia Belgica* and part of *Gallia Lugdunensis*) and the two provinces of *Germania*. Until the Augustan reorganization of 16–13 BCE, the province of *Gallia comata* extended up to the Rhine. Later, during the Principate, *Gallia Belgica*, *Germania superior*, and *Germania*

First and foremost, northern Gaul sports an impressive (and still growing) body of archaeological evidence about rural storage unique in the Western provinces (for an overview of existing data, Ferdière 2015 and this volume, with various contributions in Trément 2017; see also Huitorel 2017). No other province matches this record yet, in spite of recent or ongoing research (see Lamm and Marko 2017 for a collection of papers centered on Noricum and the surrounding regions; for Britain, Smith 2016 and Lodwick 2017 now supplement Morris 1979 but the approach remains quite superficial; the forthcoming dissertation of Vincenzo Pellegrino, in preparation at the Université Montpellier 3 and the Università di Salento, should give an overview of the situation on Spain, Southern Gaul and Italy. To the best of our knowledge and in spite of Virlouvet 2007, North Africa remains sorely understudied).

Furthermore, classification of Gallic storage facilities has been amply dealt with in previous studies and need not be tackled here (see above all Ferdière 2015). It is thus possible to move beyond a mere typological approach. For instance, existing research has shown a clear trend during the Principate towards the building of larger granaries on rural settlements. What does this mean? Are bigger storage facilities a sign of increased productivity? Or do they indicate a change in storage techniques (e.g. from bulk grain to sheaves)? This also raises the question of the impact of the Roman conquest on the rural world. The general tendency is to consider that integration in the Roman Empire resulted in a more effective and rational exploitation of resources. This is certainly true to some extent, though not all areas were affected (Leveau 2003, 2007). Furthermore, this perception is not free from the traditional narrative of Rome as a force of progress bringing civilization to a barbarian north. On the contrary, specialists of the pre-Roman Iron Age emphasize vigorously how many trends visible during the Roman period have their roots in the last three centuries BCE (Blancquaert and Malrain 2016; Reddé 2018a). Any evaluation of the impact of Rome should therefore be conducted in the *longue durée* and framed with studies of the pre-Roman to the post-Roman periods.

In this debate, one must not overlook the specific situation of Gaul within the Roman Empire. Historians readily accept that Gaul was key to the supply of the army stationed on the Rhine (e.g. Garnsey 1994): sources are clear on this point and the aim of the reorganization undertaken by Augustus in 16–13 BCE, including a census, becomes clear when it is remembered that Drusus launched his conquest of Germany in 12 BCE. That the Roman army had a massive impact on the surrounding countryside is a well-accepted fact; however,

---

inferior came under the supervision of a single fiscal procurator stationed in Trier: at least from the fiscal and economic angle, the Romans themselves thought of the whole area as one.

the economic consequences seem not to have been fully contemplated until recently (again by archaeologists rather than historians; for the Lower Rhine, see Polak and Kooistra 2013, with previous literature). But as Michel Reddé has pointed out, elaborating on a paper published by Peter Herz 25 years ago, not only was the immediate hinterland of the frontier impacted, but also the more distant areas of inner Gaul (Herz 1992; Reddé 2011b; see also Reddé in this volume). Provisioning the Roman army must have been a much heavier burden than normally envisaged. Moreover, not all the surplus grain produced in Gaul was sent to the army; towns had to be supplied too, and indeed urban *horrea* are also known in the Gallic provinces (although grain was probably not the only product stored in them; for northern Gaul most of the evidence is gathered in Guesle-Coquelet 2011). It has even been suggested, though not very conclusively, that grain from central Gaul was periodically sent to Rome, for instance during the grain shortage under Claudius (Bonacci 2015).

3 Aims and Results of the Book

Such are the problems at the heart of this book. Each contributor has addressed them from his own angle and no attempt has been made to mask diverging views. As a consequence, the reader will also find some degree of overlap between various chapters. In order to make the volume easier to read, a table containing archaeological information on granaries discussed in the various chapters has been made available online at the following URL: <https://hal.archives-ouvertes.fr/hal-01812097>. This table, compiled by Alain Ferdière with the help of Stéphane Martin, comes with a separate extended bibliography. In the following chapters, the catalogue number of included granaries is indicated as "cat. XX" when the granary is first mentioned.

Contributions have been organized in two parts. It should be borne in mind that this grouping has only been made for the sake of clarity and that there are numerous connections between papers in the two parts.

Part One deals primarily with the quantitative aspects of the problem. Can we calculate the capacity of a storage building? If so, what criteria should be taken into account, and what can we make of the numbers we come up with? Are they meaningful?

Lars Blöck is convinced that there is something to be gained from such calculations. His claims are based in particular on the well-documented case of a granary built in modern Germany, whose size was determined by the amount of products to be stored. Focusing on two well excavated villas from ancient Germany, he then attempts to estimate the size of their domains from the capacities of the storage facilities. While stressing the speculative nature of such calculations, he nevertheless finds them interesting food for thought. Given the fact that *fundi* are archaeologically invisible, his suggestions are certainly worth discussing.

On the contrary, in his contribution, Javier Salido Domínguez questions the different proposals about the storage capacity of raised granaries or *granaria* and argues for what he calls "Quantitative Scepticism". For him it is difficult to arrive at precise quantitative estimations because many factors regarding the optimum preservation conditions to be maintained inside the granaries are still unknown to us. Consequently, it is safer, in his opinion, to abstain from estimating storage capacities and to content ourselves with measurements of floor areas.

In the third paper of this section, Stéphane Martin takes a kind of middle road. Although it is clear that precise calculations of storage capacities are impossible, the author still believes it is useful to calculate plausible estimates, which can help rule out certain interpretations. On this premise, he reviews previous literature on the subject in order to build a realistic quantitative model taking into account as many parameters as possible. This model is both more specific and more general than Lars Blöck's: more specific because it does not attempt to use storage capacities to calculate cultivated areas; more general, because it aims at producing valid estimations for the whole of the area under study, while Lars Blöck stresses that his model is designed for one particular region.

The reader may find these chapters focus too narrowly on a methodological issue. However we would like to stress again how fundamental it is to ancient economic history, in particular to the current trend of quantitative studies. In absence of any textual evidence, trying to assess the capacity of storage buildings is at the moment the only way towards any attempt at quantifying agrarian production. Of course we can only aim at estimating orders of magnitude. But there is considerable room for improvement in narrowing the range of our estimates – all the more since most previous attempts at calculation storage capacities appear to have been greatly exaggerated. Our chapters argue for a more cautious approach based on a wide range of data. Indeed, to the best of our knowledge, the reader will find here the first systematic attempt at modelling storage capacities of rural granaries combining archaeological, archaeobotanical, agronomical and historical data.

Part Two offers a qualitative approach to the archaeological data, also taking into account the result of the first three chapters of the book. It is mainly concerned with explaining changes in storage methods and capacities in historical terms. Did the Roman conquest radically alter the agrarian landscape of Gaul?

How can the study of granaries contribute to this debate? Various hypotheses are proposed; they appear complementary rather than mutually exclusive.

Stanislas Bossard offers the first in-depth survey of storage facilities in Iron Age Gaul since the work of Frédéric Gransar in the early 2000s. Starting with a broad typology of the facilities attested by archaeology, the author then gives a historical overview of their use throughout the Iron Age, from the sixth to the first centuries BCE. The survey shows that structures such as souterrains, timber-built cellars or underground silos were employed in specific areas, whereas above-ground granaries were common throughout Gaul. Some farmsteads or aristocratic residences played a crucial role in storing and redistributing massive food reserves, especially cereals. It appears clear that by the time of the Gallic Wars, Gaul was already engaged in an urban and market economy, which could explain the abandonment of underground storage pits (silos) in favour of the above-ground granaries that remained the main storage facilities throughout the Roman period.

These are studied in the following contribution by Alain Ferdière. Using the conspectus of granaries in Gaul which he himself compiled (and expanded for this occasion), he shows that the increase in the size of storage facilities during the Roman period is not as widespread as generally thought. The number of rural sites with very large granaries actually appears very limited. Still there is no denying for him that there is a change in scale between the Iron Age and the Principate, indicating that the processes involved were different in nature. According to Ferdière, these changes were due to the implementation of a market economy on a large(r) scale.

The next paper by Stéphane Martin turns to the Batavian countryside. Crop production in this region lying directly south of the Rhine was always said to be of comparatively small scale due to a lack of suitable arable land. Recently some scholars have argued against this opinion, claiming that production was higher than previously thought, as demonstrated in particular by the increasing size of granaries. This increase appears in fact rather limited, but on a number of sites it cannot be denied. It is argued that these cases should be interpreted as part of a partial monumentalization of the settlements, rather than as an indicator of growing production.

In his conclusions, Michel Reddé offers a global interpretation of the evolution of the size of granaries in Gaul between the late first century BCE and the third century CE by setting the evidence discussed in the volume in a wider framework. Comparing the size and capacity of military, urban and rural storage facilities, he points out the immense amount of grain needed by the army stationed on the Rhine and the extraordinary strain producing it would have placed on the countryside of Gaul. The author then notes that large rural granaries only appeared during the second and third centuries CE. In his view, this should be explained by the eastwards shift of the Germanic frontier and the following occupation and colonization of new land in what is now southern Germany. This would have relieved Gaul from huge pressure and allowed farmers and landlords there to dispose of greater surpluses.

Clearly the case of northern Gaul is not typical for the whole of the Western provinces and one may wonder how relevant it is to a wider understanding of grain production, storage and circulation. In addition to filling a gap in current scholarship, we firmly believe it is useful to shift the light towards a non-Mediterranean region, with a different background and different dynamics. As François Lerouxel recently wrote, "*il est bon que l'histoire économique de l'Antiquité se fasse désormais à l'échelle méditerranéenne, à condition que la Méditerranée ne soit pas uniquement gréco-romaine, sinon les progrès qu'on peut en espérer ne seront pas à la hauteur des attentes*" (Lerouxel 2017, 207).

True, its core revolved around the *Mare Nostrum*; but the Empire extended far beyond the Mediterranean. Certainly the Empire was too diverse to content ourselves with a one-size-fits-all approach, yet too frequently, and more often than not unconsciously, scholars put on their Mediterranean contact-lenses and fail to see properly not only the diversity of non-Mediterranean regions, but also their importance in the making of the Empire. The north-western provinces were more than the outer periphery of the Empire surviving on military pay sent by Rome, as the first version of Keith Hopkins' model would have us believe (Hopkins 1980). This book is also a contribution to a better understanding of the economic role of the Gallic provinces through the study of the most important foodstuff of ancient times, grain. By decentring the study of grain production and trade and putting aside a strictly Mediterranean perspective, it is our hope that this volume will open new perspectives on the subject and will serve as a stepping-stone for similar studies, both on Gaul and on other regions of the Empire.

The present volume is born from the collaboration between the editor, as part of his own post-doctoral research project at the Radboud Universiteit Nijmegen within the framework of the "Anchoring Innovation" agenda of OIKOS, the Dutch National Research School in Classical Studies, and Michel Reddé, Directeur d'études at the École Pratique des Hautes Études (Paris) and principal investigator of the ERC-funded project "RurLand: Rural Landscape in north-eastern Roman Gaul".

The contributors gathered at the Institut National d'Histoire de l'Art, Paris, on 13 December 2016 to discuss a first draft of their papers during a "RurLand" workshop open to the public that left ample space for discussion. All participants are warmly thanked for the quality of the various interventions, which significantly enhanced the present volume. Special thanks are owed to Pierre Ouzoulias and Guillaume Huitorel for their thoughtful comments.

PART 1

Evaluating the Storage Capacities

CHAPTER 1

A Model for Calculating the Capacities of *horrea* and Agricultural Areas of Gallo-Roman *villae* in the Province of *Germania superior*

Lars Blöck

Compared with other pre-industrial societies, a rather high percentage of the Roman population in the Gallic and Germanic provinces was not involved in agrarian production during the High Empire. Rural produce was needed to feed soldiers and the inhabitants of *vici*, small towns and cities. To maintain this system the Gallo-Roman *villae* – the rural settlements whose economic base was the production of agrarian goods – had to produce a surplus of food and especially of grain. Although the *villae* were of fundamental importance for the Roman economy, ancient sources do not give any information about the extent of the grain production of Gallo-Roman *villae*.

Horrea, found among the buildings of several *villae*, can give us an indication about the amount of grain that was stored in a *villa* and therefore at the disposal of the owner. The problem is that there are no ancient data about the storage capacity of granaries in the Roman north-west provinces. The question of how the grain was stored in *horrea* is disputed among historians and archaeologists. Some scholars say that grain was stored in sacks, while others claim that it was stored in bulk directly on the granary floor (Rickman 1971, 85; Gentry 1976, 18; Schubert 2016, 336–38; Blöck 2016, 94 note 580). This second opinion is generally accepted. In most cases the capacity of a granary is then calculated by multiplying its inner surface area by a given height for the heap of grain. Often no explanation is offered for the height (Manning 1975, 115–18; Willems 1988, 11; Heimberg 2002–2003, 128 note 30; Schucany 2006, 197; Ferdière 2015, 39). Calculations with storage in sacks have also been put forward (Gentry 1976, 25–28; Strickland 1987; Schubert 2016, 338–47). Finally, most calculations assume that Roman *horrea* possessed only one floor for grain storage.

In my opinion, these calculations lack a sound basis and produce values for storage capacities that are far too high. Good insights into how Roman grain storage may have worked are offered by medieval and early modern granaries. Such buildings existed especially in monasteries but also in cities and on rural estates. Like their Roman counterparts, they served as storage spaces for grain

DOI:10.1163/9789004389045_003

produced in a pre-industrial (agrarian) setting. Since the late medieval and early modern granaries had more or less the same function and were of the same size as Roman *horrea*, it makes sense to take their capacity values as a reference for Roman granaries rather than using more or less random heap heights.

For early modern granaries, we possess data both on how grain was stored and on the storage capacities of granaries. So grain was to be stored in bulk in heaps no higher than 90 cm (fig. 1.1, see the transcription of the Early Modern High German text in Rudolf 1989a, 18; Krünitz 1788, 851). But the height of newly harvested grain heaps was even lower, from 30 to 60 cm (Krünitz 1788, 741, 841 and 851). This was necessary for proper ventilation of the grain in order to avoid decomposition and fermentation. Indeed, fermentation raises the grain temperature and may cause fires. From the medieval and early modern documentation on granaries, we also learn that space was needed for the delivery of the grain and for regularly moving it around (shovelling) in order to ensure proper ventilation (Krünitz 1788, 771–75, 789–90, 840–44; Johnson 1987, 164). Finally, these sources tell us that grain was stored in sacks only for transport (Krünitz 1788, 854–56).

Valuable data about the storage capacity of early modern granaries are to be found in the *Ökonomische Encyclopedie* (1773–1786) edited by Johann Georg Krünitz, where the author explains how to build such a storehouse (Krünitz 1788, 789–91). An ideal granary should measure 37.67 m by 12.56 m i.e. 473.14 m^2. The headroom should be between 2.51 m and 2.83 m. Such a granary could contain 262.69 m^3 of grain. This means that 555.21 litres of grain were stored per square metre of floor area (for the conversion of the measurements used by Krünitz into modern measurements, see Blöck 2016, 95 note 585).

Another early modern source gives us slightly different values for storage capacities, about a third higher than Krünitz's. At the end of the seventeenth century a granary, 59.34 m by 15.07 m in size, was erected in the monastery Weingarten near Lake Constance. There is an architectural sketch of this granary dated 1684 (fig. 1.1). The details of the sketch tell us that one floor with an inner surface area of 820.12 m^2 could hold 3200 Weingarten *Scheffel*. A Weingarten *Scheffel* is a measure of capacity corresponding to 183.65 l. This means that 716.58 l/m^2 of grain could be stored on one floor. The headroom of the four storeys of the granary was between 2.88 m and 3.15 m. So they were more or less the same height as the Krünitz granary (Rudolf 1989b; for the conversion of the measurements used in the early modern sketch into modern measurements, see Blöck 2016, 95 note 586).

If we take the data from the Krünitz granary and the storage capacity of the Weingarten granary respectively as the basis for estimating the capacity of Roman granaries, we get much lower storage capacities in comparison with the

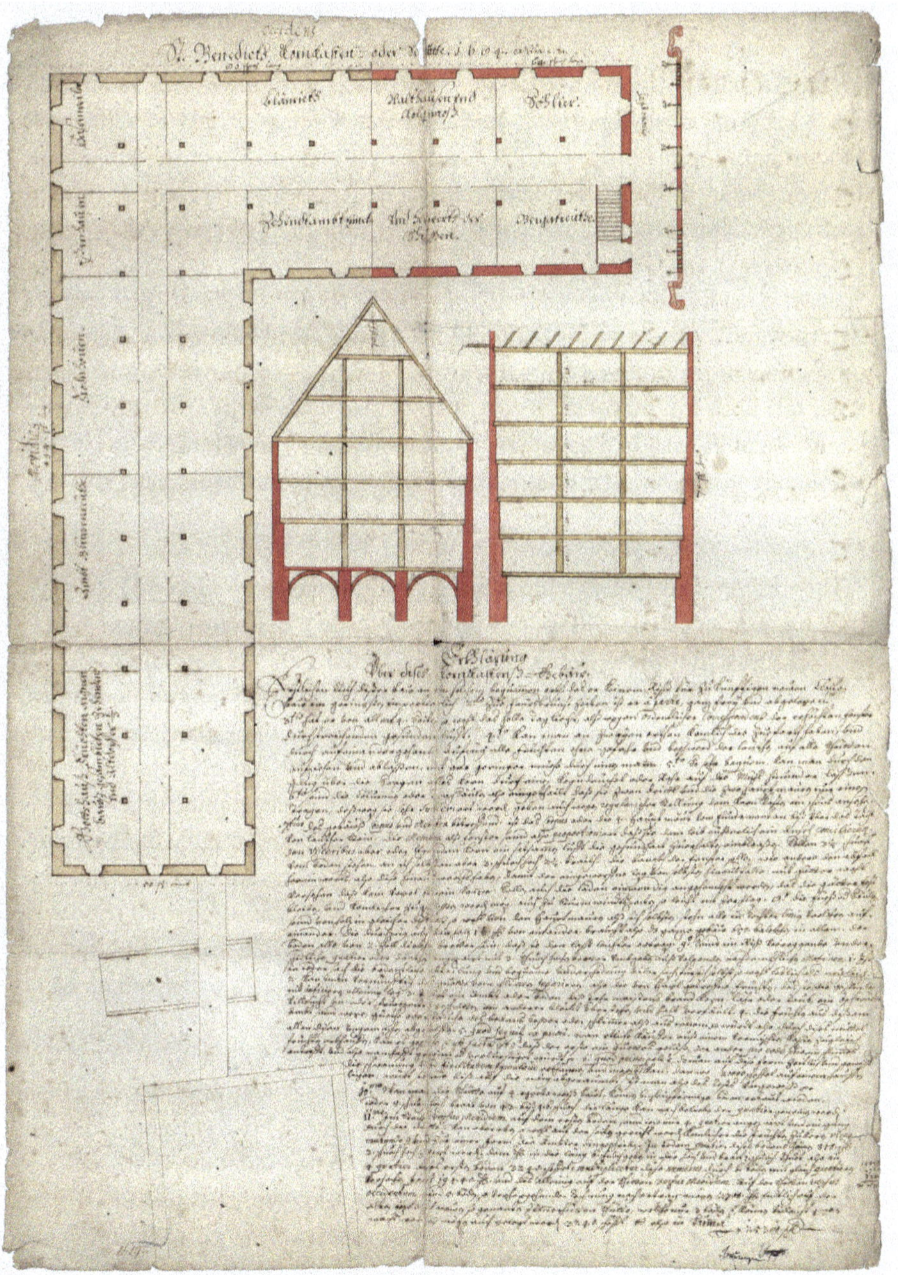

FIGURE 1.1 Architectural sketch of granary (*Fruchtkasten*) of the Weingarten Monastery from the year 1684
HAUPTSTAATSARCHIV STUTTGART, B 522 BÜ 69 PLAN; RELEASED FOR PUBLICATION AZ. 7-W 41–20170122

values published so far – assuming that Roman *horrea* only had one storage floor as proposed in the models cited at the beginning of this chapter. Thus, the storage capacity of the Voerendaal granary (cat. 33) was between 208.2 m^3 and 268.72 m^3. This corresponds to 69.4% and 89.57% respectively of Willem H. Willems' minimum storage capacity and 52.05% and 67.18% of the maximum storage capacity of the Voerendaal granary (Willems 1988). The military granaries from Roman Britain had only between 27.76% and 35.83% of the capacity presumed by Anne Gentry (1976).

But it must be asked whether Roman *horrea* actually had only one storage floor or whether some granaries might have had two. In south-west Germany, some excavations of *villae* have uncovered granaries whose walls had fallen down but with the masonry bond intact. These walls show that the Roman granaries could have had eave heights of more than 7 m (Blöck 2008, 157–59; Sommer 2005, 282–83). This means that Roman granaries could have had two storeys: a ground floor and a first floor, each with headroom corresponding to that of early modern granaries.

The existence of constructional elements such as deep foundations, very thick foundations and walls, pillars and the pilasters that most Roman granaries had, show that these were tall buildings (Blöck 2016, 96 note 590).

Horrea that did not have any substructures for a raised ground floor must have had a second floor that was used for the storage of grain. Ground floors without a raised floor were unsuitable for long-term grain storage because they were too damp and inadequately ventilated (Blöck 2016, 97 notes 592–93).

Horrea that have no substructures in the ground floor and that are well preserved show that the ground floors were not used for the storage of grain. The ground floor of the granary of the Bad Rappenau "Maueräcker" *villa* was used as a living space (Wulfmeier and Hartmann 2009, 347). On the ground floor of the double granary of the *villa* of Münzenberg-Gambach, three cellars were built, as well as a room with wickerwork walls used as a workshop (Rupp, Schmid, and Boenke 1998; Blöck 2016, 96 note 591).

Roman granaries that had substructures for a raised ground floor such as pillars or wall substructures (*suspensurae*) probably had two corn lofts: one on the raised and ventilated ground floor and one on the first floor. If one follows this reconstruction of *horrea* with *suspensurae,* the volume capacities of the military *horrea* from Roman Britain and the granary of the *villa* of Voerendaal are doubled. The granary of the Voerendaal *villa* would then have had a capacity of 416.41 m^3 or 537.44 m^3 respectively which is 138.8% or 179.15% of Willems' minimum and 104.1% or 134.36% Willems' maximum capacity (Willems 1988). The military *horrea* from Roman Britain would then have had 55.52% or 71.88% respectively of Gentry's capacity (Gentry 1976).

With the data from Krünitz and the Weingarten monastery, we can calculate the storage capacity of a Roman granary in terms of volume. Each type of grain has its own density. Therefore the storage capacity in terms of weight of a Roman granary depends on the type of cereal cultivated and stored. To calculate a Roman granary's capacity in terms of weight, we need to know which cereals were cultivated and how much of each type was stored.

The three main cereals in Roman times in the province *Germania superior* were spelt, barley, and rye. Before the industrialization of German agriculture, densities with chaff were 39 kg/hl for spelt, 58 kg/hl for (winter) barley, and 68 kg/hl for rye (Scherer 1969, 74).

The problem is that we need archaeobotanical data from a *villa* to know which cereals were stored in a Roman granary. To obtain such data, good preservation of features is necessary. But even allowing for optimum conditions during excavation, we cannot know for sure the different proportions of the stored cereals. As a consequence, to reconstruct the different proportions we must draw on more or less reliable estimates.

To model the areas necessary to cultivate the quantity of grain that can be stored in a granary, two further parameters must be known: the seed/crop yield ratio and the quantity of seed.

In the research literature on yields in ancient agriculture, these values vary greatly (Lang 2009, 393–402). Recent studies of ancient seed yield ratios presume ratios of 5:1 (five grains harvested for one sown) and 10:1 (ten harvested for one sown) respectively. But yield ratios that reach values like 10:1 are too high. Medieval and early modern sources from south-west Germany show that 10:1 yield ratios were only reached in areas with high-quality soils under special circumstances in pre-industrial grain production (Blöck 2016, 100–101 note 618). In most cases the ratio was 5:1 or even less. Since both the medieval/early modern and the Roman styles of agriculture were traditional, it is doubtful that Roman agriculture was more productive than medieval and early modern agriculture. I therefore use a yield ratio of 5:1 for my calculation.

To calculate the crop yields in the Roman north-western provinces, the existing literature often relies on information from Columella's *Agriculture* about seed quantities. But Columella writes about agriculture in Roman Italy and his values for the quantity of seed differ considerably from the values we know for medieval and early modern Germany (Lang 2009, 399–400).

The quantity of seed was stable in agriculture in Germany from medieval times until the industrial revolution in the nineteenth century (Boelcke 1964, 264; Lüning and Meurers-Balke 1980, 331). Hence, it makes more sense, in my opinion, to rely on the quantity of seed we know from medieval/early modern agriculture in Germany, particularly since the same sowing technique was used during this period and in Antiquity (Lang 2009, 397).

For the three main cereals of Roman Germany – spelt, barley, and rye – we know the following quantities of seed from late medieval/early modern sources: spelt with chaff 164.4–219.2 kg per hectare (Blöck 2016, 101 note 624), barley 147 kg per hectare (Lüning and Meurers-Balke 1980, 330 fig. 14), and rye 143.3–181 kg per hectare (Blöck 2016, 101 note 626).

As a case study, the storage capacities of the granaries of two Gallo-Roman *villae* are calculated. Based on this data, the cultivation areas are estimated that were needed to produce the quantity of grain to match the capacities of the granaries (tab. 1). Because there are two different values for storage capacities – the Krünitz storage capacity and the Weingarten granary capacity – and because we have differing values for the quantities of seed, four different values can be calculated for the storage capacity and therefore also for the cultivation areas. These values should not be interpreted as indisputable reconstructions of the cereal area but as model calculations of the maximum cultivation areas of cereals for these *villae*. It is also important to emphasize that no one calculated value has a higher probability than the other three values.

Calculations are based on assumptions that cannot be verified with the present data: the values for the quantities of seed and the seed crop yield ratios are not based on historical or archaeological sources pertaining to the research area in Roman times.

Furthermore we do not know which other crops, such as legumes and fruits, or which goods, such as amphoras or agricultural equipment, were stored in the granaries (Gentry 1976, 16; Wulfmeier and Hartmann 2009, 350). Thus we do not know how much space in the granaries was actually for the storage of grain. For example, in the *horreum* of the Biberist "Spitalhof" *villa* not only grain, but fruits, nuts, and legumes were stored, as shown by the archaeobotanical analysis (Jacomet and Petrucci-Bavaud in Schucany 2006, 590–94).

On the other hand, calculations are based on economic determinants and economic constants that can be established for pre-industrial agriculture in south-west Germany and that did not alter over the course of the centuries, such as the way grain was stored, the quantities of seed and the yield ratios. Moreover, the building and maintenance of the *horrea* were high cost factors for a Roman *villa*, as we can learn from Roman written sources on agriculture (Col. 1.4.6–7; Var., *R.* 1.13.5–6). In my opinion, it is thus more than probable that the volume of ventilated storage capacity reflects the amount of crop that was expected to be harvested.

In Roman times, grain was cultivated in a two-field system (Kreuz 1994–1995, 81; Eck 2007, 214–15; Lang 2009, 400). This means that a field was cultivated one year, and left fallow the following year. Therefore, to obtain the agricultural area of a Roman *villa*, we must double its calculated cultivation area.

The first example for the model calculation is the Bad Rappenau "Maueräcker" *villa* (Wulfmeier and Hartmann 2009; Blöck 2016, 97 note 595; cat. 251). The *villa* lies about 60 km north of Stuttgart in a fertile loess area in the Neckar valley. It had a granary with an inner surface area of 465.3 m^2. The ground floor was not ventilated and was used as living quarters as the existence of hearths shows. Therefore we must assume that the *horreum* had a first floor which served for grain storage. If we take the Krünitz capacity the granary could have stored 2583.39 hl of grain, if we take the Weingarten capacity the storage capacity was 3334.35 hl of grain.

Archaeobotanical analysis shows that rye, barley, and spelt were stored in the granary (Rösch 2009). Because all three cereals appear in the same quantities in the archaeobotanical analysis, I have hypothetically assumed that each cereal made up a third of the storage weight.

If we take the Krünitz capacity of 555.21 l/m^2 of grain, the granary could hold 1150.25 hl (44,860 kg) spelt, 773.45 hl (44,860 kg) barley, and 659.7 hl (44,860 kg) rye (tab. 1). If we take the Weingarten capacity of 716.58 l of grain per cubic metre, the granary could hold 1484.58 hl (57,898 kg) spelt, 998.24 hl (57,898 kg) barley, and 851.44 hl (57,898 kg) rye. Depending on the estimated quantities of seed, we calculate cultivation areas between 151.53 and 230.02 ha in order to produce such quantities of grain. To determine the total agricultural area of the *villa*, we have to double the area under cultivation because of the two-field system. In the model, the agricultural area of the Bad Rappenau "Maueräcker" *villa* can thus be estimated at between 303.06 and 460.04 ha.

These values seem very high but we must consider that they have been calculated with maximum grain storage capacities. Since not only grain but also other goods were stored in the *horreum*, the actual grain capacities of the granary and thus the agricultural area of the *villa* of Bad Rappenau "Maueräcker" would have been lower. There is an archaeological study of the Roman colonization of the area in which the Bad Rappenau "Maueräcker" *villa* is situated (Hüssen 2000). Even if one fits the largest agricultural area of 460.04 ha into the Roman settlement landscape there are no overlaps with other *villae* (Hüssen 2000, *Beilage* Nr. 127 and 131). So even the largest calculated area cannot be ruled out.

The second example is the *horreum* of the Biberist "Spitalhof" *villa* – a double courtyard *villa* in today's Switzerland – beside the river Aare (cat. 85). During an expansion phase in the second century, a granary was built on the estate (tab. 1). The *horreum* had a ground floor with an area of 590 m^2 that had pillars for a raised, ventilated storage floor (Schucany 2006, 187–97 and 222–23).

TABLE 1.1 *Villa* Bad Rappenau "Maueräcker" and *villa* Biberist "Spitalhof", capacity and agricultural area calculations

	Crop yield ratios 1:5 minimum quantity of seed Krünitz capacity 555,21 l/m^2	**Crop yield ratios 1:5 maximum quantity of see Krünitz capacity 555,21 l/m**
Bad Rappenau „Maueräcker"	storage capacity: 2583,39 hl:	storage capacity: 2583,39 hl
no ventilated ground floor	1150.25 hl = 44860 kg spelt	1150.25 hl = 44860 kg spelt
storage area ground floor: none	773.45 hl = 44860 kg barley	773.45 hl = 44860 kg barley
storage area first floor: 465.3 m^2	659.7 hl = 44860 kg rye	659.7 hl = 44860 kg rye
total storage area: 465.3 m^2	54.57 ha spelt fields	40.93 ha spelt fields
	61.03 ha barley fields	61.03 ha barley fields
stored grains:	62.61 ha rye fields	49.57 ha rye fields
spelt, barley, rye	= 178.21 ha cereal area	= 151.53 ha cereal area
	= 356.42 ha agricultural area	= 303.06 ha agricultural are
Biberist, Gebäude D (period II / phase 3)	storage capacity: 6551,48 hl:	storage capacity: 6551,48 hl
ventilated ground floor	3917.38 hl = 152778 kg spelt	3917.38 hl = 152778 kg spelt
storage area ground floor: 590 m^2	2634.1 hl = 152778 kg barley	2634.1 hl = 152778 kg barley
storage area first floor: 590 m^2		
total storage area: 1180 m^2	185,86 ha spelt fields	139,4 ha spelt fields
	207,86 ha barley fields	207,86 ha barley fields
stored grains:	= 393,72 ha cereal area	= 347,26 ha cereal area
spelt, barley		
	= 787.44 ha agricultural area	= 694.52 ha agricultural are

Crop yield ratios 1:5 minimum quantity of seed Weingarten capacity 716,58 l/m^2	Crop yield ratios 1:5 maximum quantity of seed Weingarten capacity 716,58 l/m^2
storage capacity: 3334,25 hl:	storage capacity: 3334,25 hl:
1484.58 hl = 57898 kg spelt	1484.58 hl = 57898 kg spelt
998.24 hl = 57898 kg barley	998.24 hl = 57898 kg barley
851.44 hl = 57898 kg rye	851.44 hl = 57898 kg rye
70.44 ha spelt fields	52.83 ha spelt fields
78.77 ha barley fields	78.77 ha barley fields
80.81 ha rye fields	63.98 ha rye fields
= 230.02 ha cereal area	= 195.58 ha cereal area
= 460,04 ha agricultural area	= 391.16 ha agricultural area
storage capacity: 8455,64 hl	storage capacity: 8455,64 hl
5055.95 hl = 197182 kg spelt	5055.95 hl = 197182 kg spelt
3399.69 hl = 197182 kg barley	3399.69 hl = 197182 kg barley
239.88 ha spelt fields	179.91 ha spelt fields
268.27 ha barley fields	268.27 ha barley fields
= 508.15 ha cereal area	= 448.18 ha cereal area
= 1016.3 ha agricultural area	= 896.36 ha agricultural area

Assuming that the granary had two storeys, it had a grain storage area of 1180 m^2. According to the archaeobotanical analysis of the granary, the cereals stored were mainly barley and spelt (Jacomet and Petrucci-Bavaud in Schucany 2006, 590–94). For the model calculation, I have assumed that barley and spelt each made up half of the storage weight. In this case the granary could hold, after Krünitz, 3917.38 hl (152,778 kg) of spelt and 2634.1 hl (152,778 kg) of barley, and after Weingarten, 5055.95 hl (197,182 kg) of spelt and 3399.69 hl (197,182 kg) of barley. Depending on the supposed quantities of seed, we get values between 347.26 ha and 508.15 ha for the cultivation area, and between 694.52 ha and 1016.3 ha for the agricultural area. The enormous storage capacity indicates that not only the estate's own harvest was stored in the *villa* of Biberist-Spitalhof, but the yield of several *villae*, which was already presumed by Caty Schucany – although on the assumption that the granary had an even higher storage capacity (Schucany 2006, 197). The phenomenon whereby one *villa* had access to the yields of other *villae* can be observed at other *villae* in the Germanic provinces. In the course of the second and third centuries there were processes of economic concentration in the rural landscape that led to increased economic and social power of some *villae* at the expense of others (Blöck 2011–2012, 108–9, 2016, 108–9).

To conclude, I would like to emphasize some crucial points. The model presented for the calculation of the storage capacity of Roman *horrea* here has many limitations as set out above. Nevertheless, it can give some idea of the amounts of grain that were at the disposal of a Roman *villa*.

The calculations of the agricultural areas are based on even more assumptions that cannot be proved. They are proposed only as tentative approaches to determining the dimensions of the agricultural areas necessary for the production of the stored grain. Furthermore, the calculated agricultural areas are not to be equated with the *fundus* of a *villa*. A *fundus* included not only fields but also woods, grassland, and other areas.

Finally, the model calculations are based on data that were extracted from sources that refer to pre-industrial agriculture in what is now Germany, and especially south-west Germany. The calculations therefore are valid for this particular region. In areas with different soil and climate conditions or different agricultural traditions appropriate datasets would be needed for such capacity calculations.

CHAPTER 2

Is It Possible to Quantify the Roman Agrarian Economy? In Favour of Quantitative Scepticism

Javier Salido Domínguez

1 Introduction

There has been a resurgence of interest in the analysis of Roman raised granaries and *horrea* in recent years. Since the publication of Rickman's reference work *Roman Granaries and Store Buildings* (1971), which started a golden decade in the study of Roman *horrea*, thanks to works by Manning (1975), Gentry (1976), Morris (1979), Petculescu (1987) and Gast & Sigaut (Gast and Sigaut 1979; Gast, Sigaut, and Bruneton-Governatori 1981; Gast et al. 1985), much has been published on this type of structure and it is currently subject to analysis by different specialists and working groups.

Research has mainly focused on the study of construction techniques, that have allowed us to correctly identify the granaries from the archaeological point of view. The publication of different typologies, the explanation of the use of certain construction materials and the application of new techniques of excavation and archaeological analysis have significantly renewed our perception and interpretation of these structures. However, for many years specialists have run up against an intractable problem: how to measure the amount of grain kept inside the warehouses. Various attempts have been made to estimate the capacities of individual granaries, usually to see if the military *horrea* would hold sufficient grain for the garrison for one year or if the rural granaries would have sufficient capacity for storing the harvest. Unfortunately the question has not been solved yet and we will try here to address this subject from the archaeological point of view.

2 Proposals about the Capacity of Roman Granaries

In the earliest published works on Roman warehouses and granaries, we already find the specialists' preoccupation with solving one of the most difficult problems in the interpretation of the archaeological data: how to calculate the capacity of raised granaries. Manning (1975, 116) in his analysis of the Roman

DOI:10.1163/9789004389045_004

military timber granaries posited that grain was stored to a height of 1.5 m. It has been suggested that if the storage was in sacks rather than in bins, the need for sloping sides to keep them away from the walls would mean that the centre of the stack would be somewhat higher than this, but this does not affect the basic calculation. In contrast, Gentry (1976, 25–28) in her study of military stone *horrea* assumed that the grain could be stored to a height of 2 m, although she already pointed out certain problems that prevent us from specifying the exact calculation. Based on this and on the granary's floor area, Gentry proposed in her work (1976, table 2) the capacity in m^3 and the theoretical annual grain-holding capacity of different *horrea*. In the opinion of Sigaut (1981, 168), for a pile about 2 m wide at its base, it would be impossible to store the grain to a height of more than 1 m. More recent archaeobotanical studies, such as the analysis carried out in the urban warehouse of Amiens (FR), have determined that the average thickness of the cereal layer should be between 20 and 40 cm (Matterne, Yvinec, and Gemehl 1998; cat. 250) to maintain optimal conservation conditions of the cereal and avoid germination. This proposed height has been accepted by various specialists for calculating the capacity of certain granaries such as the *horreum* of the *villa* of Emptinne-Champion at Hamois (BE) (Van Ossel and Defgnée 2001; cat. 05 & 47), although it is also true that the total amount of grain stored at this height appears small.

From these initial proposals, calculations of capacity proposed by archaeologists have been numerous and varied. Each specialist generally presents a different storage height for the cereal layer or different weights of the grain in each approach. In other cases, minimum capacities have been proposed without taking into account the resistance of the structures to support the weight of the cereal. In two monographs I have already indicated some of the most interesting studies that aim to determine the volume of cereal that could be stored in the *horrea*, both in military and rural settings (Salido Domínguez 2011, 53–55, 2017, 81–82). I do not intend in this paper to point to more studies that try to justify the published capacity calculations, but only to indicate certain factors that allow us to question some theories and to conclude that these attempts are both inaccurate and incorrect.

3 The Contribution of Archaeobotanical Analysis to the Study of the Capacity of Roman Granaries

Archaeobotanical analyses are nowadays one of our main sources of knowledge of ancient agriculture, the introduction of new crops, the use and extension of forests, ancient diet, as well as redistribution and long-distance movement of

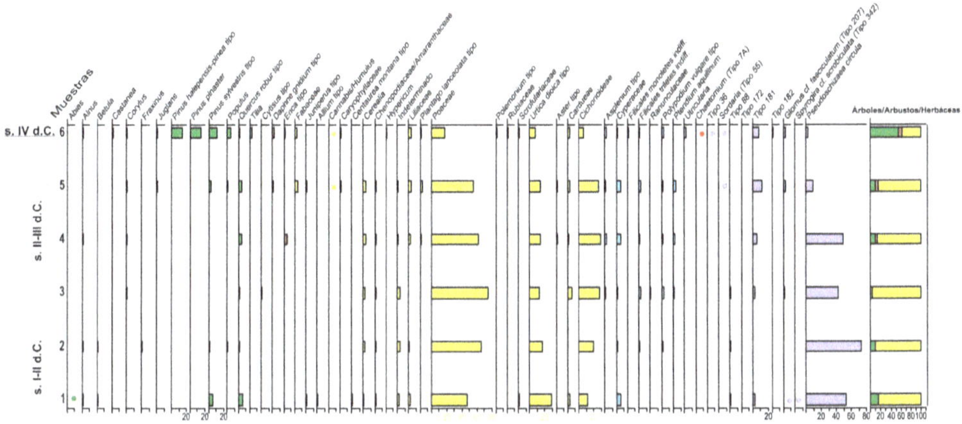

FIGURE 2.1 The palynoligical results on pollen samples from the Roman granary of Veranes (Gijón, ES). Column E5
FERNÁNDEZ OCHOA ET AL. 2012, FIG. 74

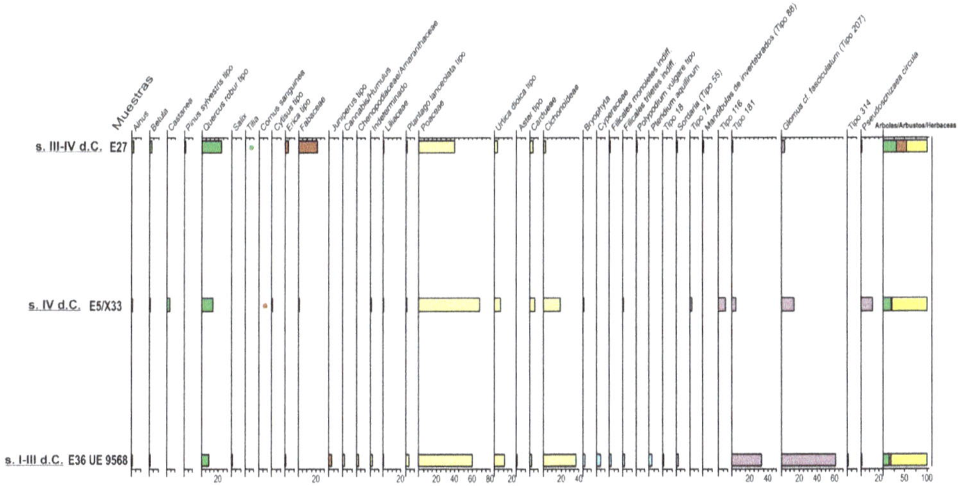

FIGURE 2.2 The palynoligical results on pollen samples from the Roman granary of Veranes (Gijón, ES). Column E27, E36 and room E5
FERNÁNDEZ OCHOA ET AL. 2012, FIG. 74

foodstuffs, and trade in Roman times and, of course, for the study of granaries and their storage capacity. Although ever more effort is being made to approach the production and storage of food from the study of archaeobotanical samples, they are still to be fully integrated into our analysis of the Roman economy. The development of these analyses allows us to interpret as granaries certain buildings for which, because of their state of preservation, construction techniques offer little information.

At certain sites, only palynological analysis can give us data about the type of agricultural production and the environment in which the granary is located. This is the case of the *horreum* of the Roman *villa* of Veranes (Gijón, ES) (Fernández Ochoa et al. 2012, 161–70). When it was used as a granary in the early Roman Empire, the surrounding landscape was heavily deforested and left practically treeless (Figs. 2.1 and 2.2). Archaeobotanical analyses by J. A. López confirm that there were only deciduous trees, scrub, and bushes in the *horreum*'s surroundings during the early Roman Empire (i.e. English oak/*Quercus robur*, alder/*Alnus*, birch/*Betula*, ash/*Fraxinus*, hazel/*Corylus*, poplar/*Populus*, juniper/*Juniperus*, tree heath/*Erica arborea*, broom/*Cytisus*, and walnut tree/*Juglans*). On the contrary, during the Late Empire, when the *horreum* became more of a pantry and less of a warehouse, the *villa*'s surroundings were slowly reforested through the recovery of deciduous trees and bushes as well as the repopulation of the forest with various tree species such as the maritime or cluster pine (*Pinus pinaster*), walnut (*Juglans*), and chestnut (*Castanea*). Similarly, this period witnessed a substantial decline in pollen grains (*Cerealia*) due to progressive reforestation and the lack of agricultural activities in the Veranes surroundings in the fourth century CE. As a result, the research on construction techniques and the archaeobotanical data both confirm a functional change in the original *horreum* during the late Roman Empire.

The interpretation of the data provided by the palynological analysis together with the macro-remains that give us more information about the study of agricultural practices are still meagre and the conclusions that we can reach are certainly partial. Nevertheless, we will try to approach the archaeobotanical problem in relation to the study of Roman granaries. The main problem that we find is the absence of the application of these analyses to the study of the granaries in some regions, being practically non-existent for the moment in the Mediterranean and very scarce in the case of the British Isles. We also find little detailed information about the stored wheat species or the mode of storage of cereals, without determining if they were kept as spikes, spikelets or grain, or if insects or signs of germination were present.

3.1 *What Species of Grain were Stored in the Roman* horrea?

The cereal species stored in Roman rural granaries were varied. In some *horrea* such as the granary of Maurepas (FR) (Coulon 1987; Marinval 1993), only naked wheat was found, while in three other granaries, naked wheat was documented together with hulled grain as in the *cella penaria* of Bliesbruck (FR) where spelt (*Triticum spelta*) was found (Petit 2005), in the granary of Borg (DE) in which barley (*Hordeum vulgare vulgare*) and emmer wheat (*Triticum dicoccum*) were recognized (Wiethold 2000) or inside the granary of Parsac "Laschamps" (FR), in which spelt and emmer wheat, as well as oats were documented (Roger and Bet 2005). Naked and hulled wheats were also found together in a number of other granaries, such as the *horreum* of the Roman *villae* of Wendens Ambo (Hodder 1982), Bad Dürkheim "Ungstein" (Piening 1988), Bad Rappenau "Babstadt" (Wamser 1977; Wagner-Roser 1987; Herberg 1994, 1998, 1999; Hüssen 2000; Hartmann and Meyer 2001; Hartmann and Reichle 2002), Emptinne-Champion (Van Ossel and Defgnée 2001), Houdan (Philippe 1993; Philippe et al. 1994; Matterne 2001, 155), and Maasbracht (Kooistra 1996, 253–76). When considering the presence of common and durum wheat (*Triticum aestivum* and *durum*) in the *horrea* of Roman *villae* it should be remembered that they can be used in bread-making, which suggests that the grain preserved in the granaries of Roman rural centres was for local consumption.

3.2 *Not Just Grain: Other Foodstuffs Stored in the* horrea

In some warehouses, different species of wheat and other foodstuffs, such as legumes, were conserved possibly in different storage bins. Legumes were stored increasingly in the Roman era and they could be conserved together with cereals. This is the case of lentils preserved with barley in the granary of Marolles-sur-Seine (FR) (Matterne 2001, 155) or in the *horreum* from the *agglomération secondaire* of Crest; in another store building of Crest naked wheat, barley, beans, and fruits were all stored together (Bastard and Stephenson 2002). There is also evidence of other legumes such as peas in Jülich "Kirchberg, *WW 112*" (DE) (Jenter and Wippern 1997; Päffgen 2000, 283–84; cat. 106), wheat, oats, barley, peas, and beans in Alle (CH) (Othenin-Girard 1999, 2002; Othenin-Girard and Elyaqtine 2002; Demarez et al. 2010; cat. 125 and 221) and Villiers-le-Sec (FR) (Matterne 2001, 155; Wabont, Abert, and Vermeersch 2006, 440–42), and other types of wheat together with peas and beans in Voerendaal (NL) (Willems 1986a, 1987a, 1987b, 1992, 1995; Willems and Kooistra 1987, 1988; Kooistra 1996; Matterne 2001, 158; cat. 33).

The wide range of crops characteristic of the Roman period is also reflected in the archaeobotanical findings of species that must have been introduced or imported, possibly in relation to arboriculture and trade (Willcox 1977). Production of these crops probably boomed in classical times. Fruit species could be transported over great distances using different techniques of preservation, in dried form (figs, grapes, plums) or in *amphorae* with liquids (olives, peaches). In Biberist (CH), besides barley and spelt wheat, remains of fruits such as plums, cherries, etc. were found (Schucany 2006; cat. 85). Carbonized grapes were found in the warehouse of Lournand (FR) (Devauges 1979, 457–58; cat. 132).

3.3 *How was the Grain Stored?*

During the storage period, it was necessary to air the grain properly, regularly inspect the stocks, and clean the grain stored in order to keep the cereal dry at a uniform temperature and protected from insects, rodents, and other pests (Fig. 2.3). These requirements were of great importance to ensure safe storage. In raised granaries, the commonest way to store grain was to pour it in bulk or in spikes directly onto the floor, whether they were naked or hulled cereals. Archaeological data show that cereals were stored in bulk in the military stone granaries of South Shields (van der Veen 1988), Valkenburg (Pals, Beemster, and Noordam 1989), as well as in the urban areas of Alcester (Booth 1989), London (Straker 1984), Amiens "ZAC Cathédrale" and in the granary of the roman *villa* of Voerendaal (Kooistra 1996). Classical sources mention the construction of spaces or bins to separate the different types of grains (Col. 1.6.9–24) and the study of the archaeobotanical remains found inside the *horreum* of Amiens "ZAC Cathédrale" confirms some internal organization and separation of products in the storage chamber (Matterne, Yvinec, and Gemehl 1998).

Conservation of the grain in spikelets improved the protection against the risks associated with humidity, often responsible for early germination or insect damage. Grain husks also ensured the cereals remained clean throughout the storage period. To that end, Pliny (*Nat.* 18.73) recommends storing grain in spikes. They were mostly stored with glumes in the granary of Houdan (FR). Although few analyses have been made, it is possible to determine that hulled grains were often stored in spikes or spikelets after initial sieving or threshing. In contrast, naked grain was usually stored clean in Roman rural centres.

In point of fact, despite the advantages of storing the cereal in spikelets that better isolates the grain from moisture and reduces the risks of germination, it was also stored completely clean in Roman times. The fact that grain was stored dehusked could indicate that it was to be consumed on the spot. This could in turn indicate a high degree of self-consumption, meaning that grain would have been consumed where it was produced. This was the case of the granary of the Roman *villa* of Wendens Ambo, where a grain dryer contained common, spelt, and emmer wheat (*Triticum aestivum*, *spelta*, and *diccocum*).

Some studies have documented the weeds found with the grain inside the warehouses. In the granary of the Roman *villa* of Champion, bromine (*Bromus sp.*), cleavers (*Galium aparine*), *Polygonum Convolvulus*, pale persicaria (*Polygonum lapathifolium*), *Chenopodium album*, *Plantago lanceolata*, and *Rumex acetosella* are found. Also interesting is the presence of alfalfa (or lucerne,

FIGURE 2.3 Reconstruction of the *tabulatum* and supports of the granary of Veranes (Gijón, ES)
FERNÁNDEZ OCHOA ET AL. 2012, FIG. 57

Medicago sativa) and cultivated pea (*Vicia sativa*) that were grown as fodder (Van Ossel and Defgnée 2001). *Medicago lupulina* and wild radish (*Raphanus raphanistrum*) were also detected among the different types of wheat found in the granary of Maasbracht. In the *horreum* of Voerendaal, the grain was threshed and cleaned before being stored, but there was a high proportion of weeds and straw with common corn-cockle (*Agrostemma githago*), rye brome (*Bromus secalinus*), and Venus' comb (*Scandix Pecten-veneris*) being predominant. The grains of common windgrass (*Apera spica-venti*) and the seeds of poppy (*Papaver dubium/p. Rhoeas*) attested in this granary were also remains of threshing.

3.4 *The Consequences of Poor Environmental Conditions or Prolonged Storage: Insects*

Grain was exposed to a number of pest insects that could cause a major decline in cereal quality. The problems of germination of stored grain increased considerably in Roman times with the construction of huge store buildings while the conservation of large volumes of grain for extended periods for market distribution and long-distance transport posed major problems and entailed losses. Germination might either be controlled for malting and brewing or be accidental, caused by adverse environmental conditions with excessive relative humidity (20–25% or more). Controlled germination has been found in the granary of Bliesbruck where spelt wheat was documented still wrapped in its glumes. It was beginning to germinate at the time of its carbonization, suggesting that it was used for the production of beer (Petit 2005). It is important to remember that the conservation of hulled rather than dehusked grain increased protection against a damp environment, which was often responsible for early germination or degradation by insects. As in Amiens "ZAC Cathédrale", much of the wheat found in a cellar of Mont Beuvray (FR) was also burned when it had germinated (Barral, Beck, and Bonenfant 1993; Barral et al. 1995). It is probable that germination was due not only to the conservation conditions, but also to the duration of storage.

We do not know how long the cereal was kept in the granary. If we consider the textual information, Ausonius boasts about storing harvests for two years in his *villa* (*Hered.* 27). On the basis of this information, it would be useful to ask whether the granaries served for storage of all or part of the harvest. If they could contain all the reserves collected, the study of their capacity would even allow us to determine the area of land suitable for cultivation. Thus, we could estimate the hectares of the *fundus* taking into account the areas reserved for livestock, the forest mass, and the area under cultivation. In this regard, I believe there are some data that allow us to question this theory and to argue that many granaries did not conserve all the crop. Most of it was probably marketed and not stored in the rural granaries (Cato, *Agr.* 2.7). The *Digest* indicates that the preserved provisions were "more available for maintenance than for cultivation" (*Dig.* 33.7.12), so the cereal harvested seemed to be intended for these two purposes. On the other hand, we should take into account that the interest of the *dominus* lay in selling agricultural surpluses, rather than accumulating products, because these could be lost due to various factors (humidity, temperature, insects, rodents, etc.). It should also be borne in mind that damage caused by long-term storage might not only result in the loss of the cereal, but also adversely affect the quality and the nutritive value of the product. In this market logic, it seems more convenient to send the harvest for sale in the urban areas (Salido Domínguez 2013; Salido Domínguez and Bustamante Álvarez 2014). This would explain why most of the granaries were of medium dimensions, up to 10 m long and 5 m wide (Salido Domínguez 2017, 88). In conclusion, most of the harvest was not intended for storage, but for sale and, in consequence, the stored cereal should not be taken as representative of the agricultural capacity of a Roman rural centre. In a military context, Tacitus (*Agr.*, 12.2) indicates that the granaries had enough capacity for one year. However, it was in the Roman army's interest to store it and so have grain for the winter (6 months). According to Gentry's calculations, with a storage height of 2 m, there was enough space to store grain for at least two years if not more. Since there is no doubt that these calculations are incorrect, these proposals should be challenged.

4 Quantitative Scepticism

Several factors enable us to question the proposed quantification of Roman granaries' storage capacity, which lead to what we call the "Quantitative Scepticism":

1. We do not know what type of cereal was stored in most of the granaries (see section 3.1). This information is essential, since the weight of each species of cereal is different (Gentry 1976, 25).
2. Not only grain but other products were preserved in the *horrea* as made clear by archaeobotanical analysis (see section 3.2).
3. We generally do not know the mode of storage: in bulk, in sacks, or in bins and if it was kept in spikes or cleaned (see section 3.3).
4. A theoretical calculation can be proposed for a full granary, but these were continuously emptied at each harvest according to the needs of their owners. A *horreum* will never be able to provide us with an estimate of the agricultural potential of a rural centre, since most of the harvest was destined for sale, redistribution, and almost immediate commercialization. The owner of a rural establishment was not economically interested in keeping a product as perishable as grain (see section 3.4).
5. Archaeobotanical analyses indicate that not all the surface of the floors would be covered by grain, because it was necessary to leave free space for circulation and ventilation to prevent the grain from germinating. In addition it is very possible that there were different compartments in the interior to separate the types of grain (see section 3.3).
6. It is necessary to calculate the load-bearing capacity of the *tabulatum* or floor of the granary, which is hard to do unless we know the construction material. Most of the granaries were made of wood, but the floor strength also depends on the species of wood.
7. The research experts do not agree about the right height of the pile of grain to ensure it did not spoil (see section 2). Undoubtedly, the height of the pile would depend on the time of storage and the possibility of accessing the different parts of the storage chamber to remove the grain and thus prevent germination or infestation of the lower part of the heap where higher temperatures reigned.

8. We do not know how long grain was kept in the granary. If we consider the textual and archaeological information, we can conclude that the stored cereal was not representative of the agricultural capacity of a Roman rural centre which would not necessarily keep the grain for a year (see section 3.4).

Conclusion

Information from archaeobotany and ethnoarchaeology warns us of the need to review the studies of quantification of the maximum grain capacity stored in Roman *horrea*, to reconsider the data referring to the agricultural exploitation of a territory, as well as to the quantity of cereal estimated for direct consumption and export. In short, we propose to question or relativize the proposed capacity, until multidisciplinary studies combining the physical principles of the load-bearing capacity of the *tabulata* and of the preservation of the grain allow us to address this problem in the most appropriate way possible.

In my view there is no need to publish further quantification models of the capacities of Roman granaries. Therefore, I advocate taking into account the qualitative rather than the quantitative aspects of the storage in *horrea* to compare and contrast the capacities of the different granaries. In this way, comparison of the volumes of the different *horrea* in order to classify the buildings according to their storage capacity could help determine the centres of production, intake and provisioning and allow us to reach conclusions about the procedures for collecting and shipping grain, and the commercial networks and techniques for supplying grain in Roman times. We hope that physical studies, with the help of new technologies, will enable us to solve in part this archaeological problem in the near future.

CHAPTER 3

Calculating the Storage Capacities of Granaries: a Tentative Model

Stéphane Martin

The previous two chapters present differing views on the feasibility and the utility of calculating storage capacities. There is no doubt that exact calculations will always elude us, for lack of proper sources. In his contribution, Lars Blöck insists rightly on the modelling aspect of his calculations. Notwithstanding the justified scepticism of Javier Salido Domínguez, it remains true that even the maximum capacities estimated by Lars Blöck lead to a serious reappraisal of previously published estimates. However, Blöck himself stresses that his model is designed for the area corresponding to modern south-west Germany and that it should not be applied as it stands to other regions. The aim of this chapter is to present and study in greater detail the various parameters necessary for calculating storage capacities, in order to set out what could be considered an average situation, valid for the whole of the research area, and possibly beyond.

In the north-western provinces, airtight storage in underground silos is not attested during the Roman period: it disappeared at the end of the Iron Age (La Tène D) until the Early Middle Ages (see Chapter 4 for the Iron Age, and various contributions in Vigil-Escalera, Bianchi, and Quirós Castillo 2013 for the Middle Ages). Calculating the storage capacities of an underground silo is fairly easy, since grain will be preserved only if the silo is completely filled before being sealed. Underground silos thus offer one of the rare cases when the total volume of a structure equals it operating volume. However, it should be noted that for Frédéric Gransar, Iron Age silos were used to store grain in spikes or spikelets, whereas for François Sigaut, silos are mainly used for grain that was already threshed and winnowed (Sigaut 1988; Gransar 2003). Sigaut does mention underground storage of spikelets, but as an uncommon phenomenon. Sigaut used mainly historical and ethnographical documentation, whereas Gransar used archaeological evidence: this may well be the main reason for their differing interpretations. This matter is nonetheless worth exploring further, since chaff can represent 10 to 50% of the total weight of hulled grain (Ouzoulias 2006, 174. For emmer and spelt, see also Saunders 1904 who gives a lower proportion of chaff than Ouzoulias).

DOI:10.1163/9789004389045_005

As evidenced by clear archaeological data, the shape and size of granaries from the Roman period differed from their Iron Age predecessors (see Part II of this volume). In most regions, wooden granaries on posts were gradually replaced, from the first century CE onwards, by larger, stone-built structures. However, the underlying technical principles of storage remained unaltered: grain was still preserved through control of the atmosphere by ventilation (to use the categories devised by Sigaut 1988, 17). The following discussion therefore concerns all granaries, regardless of their architectural type and their chronology. There is likewise no technical difference between a small rural granary and a big urban or military structure. For this reason, it is legitimate, in my view, to use data from non-rural contexts to understand storage in the countryside and vice versa. However, due consideration must be given to at least two factors. First, the function of the storage structure under study in the chain between harvest and supply. This chapter will focus on storage in a rural context, in most cases on production sites. What is at stake is therefore likely to be mid- to long-term storage, either for on-site use and consumption, or for sale on the market at a later date, with good conservation as the main goal. Second, our reconstructions must naturally take into account all available archaeological data; primary evidence, however meagre, should guide our use of historical and ethnographical comparisons, and should take precedence over evocative but sometimes ill-founded parallels (see Halstead 1987, with examples from the Mediterranean).

In trying to assess the storage capacities of granaries, scholars have taken two approaches. The older attempts took their cue from the buildings' architecture, generally taking storage in bulk as a given. Thus Francis Haverfield and Robin Collingwood for wooden, and Anne Gentry for stone military granaries in Britain (Haverfield and Collingwood 1920; Gentry 1976). The first two authors thought that grain was stored in bulk, stacked in bins against the walls to an average height of 6 feet (ca. 1.80 m) on both sides of a central corridor. Gentry calculated that the walls of the Corbridge granary could stand the pressure of a grain heap of ca. 3 m (but this was not her favoured solution, as she thought grain was stored in sacks).[1] More recently, Emanuele Papi and Francesco Martorella (2007) used the same method to reconstruct heaps as high as 4.90 m in a granary in *Thamusida* (in modern Morocco)![2] Without using such calculations, Gustav Hermansen (1982, 226–37) combined evidence from present-day Canada, where a storage height of 8 feet (ca. 2.45 m) is common, with the

1 With Richmond and McIntyre 1938–1939, these works have been the basis of all subsequent calculations for Roman military granaries, listed in Salido Domínguez 2011, 53.

2 Even if this granary is considered as a transit warehouse, i.e. for short-term storage before shipment, as is the case of Federico De Romanis (2007), such a height is surely excessive.

height of the storerooms in *Ostia* and *Portus* near Rome, to suggest 3 and 4 m heights for grain heaps in *horrea* from each site (he does acknowledge that these are maximum values, and that it is rather impractical to store grain in heaps higher than 2 m).

This approach, which does not seem to have been applied to rural buildings, either does not give due consideration to the conditions necessary for grain preservation,[3] or, in the case of Hermansen, rests rather uncritically on recent parallels. Funnily enough, the thickness of granary walls does not seem to be interpreted in terms of thermal insulation, although this was well known to an eighteenth-century writer such as the abbot Tessier: "*Les murs doivent être de bonne épaisseur, pour garantir les blés de l'humidité & de la chaleur*" (Tessier 1793, 460; Plin. *Nat.* 18.300 is rather ambiguous but might be pointing to something similar).

The second method starts from the nature of the stored produce rather than from the building. This appears wiser since, as Sigaut wrote (e.g. 1988, 18–19), storage structures are generally built according to the form of the produce. It is clear from various kinds of evidence that a single granary could host a variety of products, either different types of grain as for example in Amiens "ZAC Cathédrale" and Tiel "Passewaaij", granary 47, or different types of produce, e.g. grain and legumes in Alle (Matterne, Yvinec, and Gemehl 1998, here cat. 250; Kooistra and Heeren 2007, here cat. 312; Demarez et al. 2010, here cat. 125 and 221). Furthermore, not everything might be for human consumption, although it can be difficult to distinguish between food and fodder (Jones 1996). However, much of the archaeological evidence does relate to the storage of grain and in the following pages, the focus will be on wheat and barley, the main crops for the period and region under study.

The calculation of storage capacities needs to take four main parameters into account. The first one is obviously which crop is stored, since density varies from species to species. The second parameter is closely linked to the first: in what form is grain stored, i.e. how advanced is the processing of grain when it is brought into storage?

Sigaut (1988, 6) distinguished among several forms of grain storage. Sheaves are notoriously hard to find in the archaeological record, all the more so because they can be stored in stacks and leave no tangible traces. Some buildings have been identified as barns but only on typological grounds (Ferdière 1985). We do have hard evidence for storage of grain in bulk and grain in spikes or spikelets. The difference between naked or dehusked grain on the one hand,

3 Although he also recreates heaps resting against the walls of the *Grandi Horrea* in Ostia, Nicolas Monteix, in Boetto et al. 2016, 213–20, does take these parameters into account, which leads him to propose a much more reasonable 1.60 m as a likely height.

and hulled grain on the other, is key to understanding the nature of storage. Lars Blöck notes that finds of hulled grain appear more frequent in rural contexts, while dehusked grain is more common in urban and military settings (Blöck 2011–2012, 93–4). This is logical since grain stored hulled or mixed with chaff can be preserved for a longer period, as was already known to ancient authors (Plin. *Nat.* 18.306; see also Var. *R.* 1.72). Various estimates of grain density have been put forward; further research is needed but for the region under study, the data compiled by Pierre Ouzoulias will be used here, as they appear both consistent with ancient sources and with proposals by other authors (Ouzoulias 2006, 173–76. For emmer and spelt, Saunders 1904 and Stallknecht, Gilbertson, and Ranney 1996, both using North American data, give slightly lower values. But figures given in Kooistra 1996, 98 and 318, appear decidedly too low: see Chapter 6, this volume).

As a general rule, we should assume, for our calculations, storage in spikelets or with chaff, rather than threshed and winnowed grain. Indeed, in the small urban granary of Amiens "ZAC Cathédrale", the main crop, spelt wheat, was stored hulled (Matterne, Yvinec, and Gemehl 1998). If this building was the storage facility of a merchant, as hypothesized by the authors, this means he bought the grain unprocessed from the producer. The merchant would have had the grain threshed and winnowed later, before sale: it is clear from the Edict of Maximum Prices that selling fully processed barley was much more profitable (100 denarii per *modius* instead of 30 for husked grain: *Edict. Diocl.* 1.7).[4] However we should not completely rule out the possibility that grain was processed and sold directly by the producer (or the owner of the estate). The rare discovery of a second-century CE shipwreck with its cargo in Woerden (NL) confirms that grain could be shipped in bulk, threshed and winnowed (Pals and Hakbijl 1992; see Haalebos 1996 for the archaeological context). The grain appears to have been stored for some time, already threshed and winnowed, before shipment: there is no way to decide whether this storage took place on the production site or elsewhere.

The third parameter to consider is the usable surface area of the granary. The need for free space is regularly mentioned in agronomic literature, at least since Olivier de Serres (1600, 135), but although Haverfield and Collingwood took it into account in 1920, it does not appear in all estimates of storage capacity. The necessary amount of free space is conditioned by the form of storage.

There are three ways of storing grain in granaries: in bulk, in bins or in sacks (to my knowledge, storage in large jars, as known in the Mediterranean, is not attested in north-western Europe). Storage in bins is now rejected by all

4 Reference kindly supplied by Pierre Ouzoulias.

scholars for one good reason: there is no archaeological evidence for it, even in the best preserved granaries (Rickman 1971, 85; Gentry 1976, 18).

Storage in bags is favoured by some, most notably Geoffrey Rickman and Anne Gentry, followed by Anne Johnson (1983) and more recently by Tobias Schubert (2016, 336–37) in his dissertation on rural granaries from the Lower Rhine.[5] In all cases, the authors consider the grain naked or threshed and winnowed. The first two had a very rational approach, underlining the practicalities of storage in sacks for handling (Rickman 1971, 85–86; Gentry 1976, 18–22). But one can agree with Catherine Virlouvet and Nicolas Monteix when, discussing Rickman, they write that although his ideas look good on paper, he underplays the constant care needed by grain unless it is very dry (Virlouvet 2015, 680; Monteix in Boetto et al. 2016, 213–14). Furthermore, the presence of *saccarii* near *horrea* does not mean that goods were necessarily carried in bags, since the term apparently designated all kinds of porters (Virlouvet 2015, 676–77). Schubert (2016, 336) adds one piece of archaeological evidence in favour of storage in sacks. In 1818, a fourth-century CE *burgus* was excavated in Engers (near Koblenz, DE). The evidence points to a violent destruction by fire already in ancient times, which led to the charring of large quantities of grain, mixed with melted lead. The presence of this lead, as well, according to Schubert, as the mixture of various kinds of grain, is for him evidence that grain was stored in bags. It is worth citing here the original report as published in 1826:

Ausser den römischen Gussmauern und sonstigen Baumaterialien gehören zu den umbezweifelt achten, hier vorgefundenen Anticaglien, die Scherben von terra cotta, Gefässstücke aus grobem Thon; so wie die sehr grosse Masse verbrannten Getraides, welches in Schichten von 3 bis 9 Zoll Dicke [ca. 8 to 24 cm], *an einigen Stellen sogar mehrere Fuss hoch lag* [in this case, 1 foot = ca. 29 cm]*; es war Roggen* [rye], *Gerste* [barley], *meistens aber Weizen* [wheat], *wahrscheinlich also ein Getraide-Magazin gewesen. Die Menge grosser Stücke tropfsteinartig geschmolzenen Bleies, die sich in dem Getraide vorfanden, könnten zu der Vermuthung fuhren, der Thurm sei mit Bleiplatten gedeckt gewesen* (Dorow 1826, 24).

First of all, nowhere is it written that the various kinds of grain were mixed. Second, although it is written that lead was indeed found in the grain, it is not clear why this should indicate storage in bags, all the more since Schubert accepts Dorow's interpretation that it came from the roof. Third, 8 to 24 cm layers are consistent with storage in bulk; heaps several feet high less so, but the original report seems to indicate that this was not the general situation (see below for a discussion of the height of the heap). Based on comparison with similar

5 One can also cite Donald Strickland (1987) whose approach is less satisfactory than the others: storage in sacks is taken for granted because it is modern practice, and it is not specified whether the grain is hulled or not.

buildings, there was probably at least one upper-floor: could the grain have fallen during or after the fire (see the better preserved *burgus* in Zullestein: Baatz in Reddé et al. 2006, 227–28, *s.v.* Biblis)? Most importantly, the type of site must be taken into consideration: Engers is a late-antique fortified post interpreted as a harbour or a landing station, probably built during Valentinian's reorganization of the Rhine frontier. Can the storage method adopted in this context be applied to earlier civilian contexts?

This does not mean that storage in bags did not exist; during the excavation of Amiens "ZAC Cathédrale", some textile fibres were recovered among the grain, possibly indicating sacks (Matterne, Yvinec, and Gemehl 1998, 106). But it should be remembered that we are dealing here with *hulled* spelt, and not naked or dehusked grain, stored in a probable merchant warehouse, and not on a production site. As a general rule, sacks are only adapted to short-term storage (Geraci and Marin 2016, 88).

All in all, storage in bulk, whether the grain was fully processed or still in spikelets, is the most likely solution. For what this testimony is worth, it should be noted that storage of hulled grain on the floor is described in the peasant-setting of the *Moretum*, a short poem formerly attributed to Vergil (*Mor.* 16).[6] It is thus supported by archaeological and textual evidence, as well as historical and ethnographical parallels.

In military granaries, or more generally large rectangular granaries, scholars normally assume a central corridor at least as wide as the entrance door, with grain stacked in heaps against the walls. Haverfield and Collingwood, Gentry, Papi and Martorella, and Monteix, concord on this point. Yet from the eighteenth century onwards, agronomists insist on the need to keep some free space between the walls and the heaps, varying from ca. 30 to 120 cm (1 to 4 feet) (Table 3.1). In addition to providing room to move around, this was to avoid both dust and pests falling from the walls, and grain falling in the gaps between the wall and the floor (particularly if it was made of boards).

To this should be added the space necessary for accessing the granary and walking around the heaps, whether a single central corridor or smaller passageways. Finally, extra space is needed if grain is stored threshed and winnowed, because it needs frequent shovelling. The operation consists in moving the grain from one place to another by tossing it in the air with a shovel (hence the name). It should be done very frequently in the first six months after the harvest, with decreasing frequency until the end of the first or second year (authors vary). The problem with shovelling in Antiquity is that ancient authors

6 The evidence of the archive of the *Sulpicii*, a family of tradesmen from *Puteoli*, also seems to point to storage in bulk: *TPSulp.* 45 and 46, commented on with differing conclusions by Virlouvet 2000 and Jakab 2014.

TABLE 3.1 Free space between heaps of grain and walls according to various authors

Publication	Free space
Duhamel 1768	ca. 60 cm between wall and grain
Duhamel 1768	In the Grenier d'Abondance in Lyon, ca. 120 cm between wall and grain
Krünitz 1788	ca. 60 cm between wall and grain
Diffloth 1907	40 to 50 cm between wall and grain, and a central passage at least 100 cm wide
Haverfield and Collingwood 1920	Central corridor of ca. 90 cm (theoretical)
Richmond and McIntyre 1938–1938	Central corridor of ca. 300 cm (theoretical)
Demarez et al. 2010	100 cm around heap of grain (theoretical)

apparently do not mention it at all; textual evidence of this practice is first attested in the sixteenth century (Beutler 1981, 37–39). This led Sigaut (1988, 8) to think that it had not become common practice until the eighteenth century. However, it is hard to conceive storage of threshed and winnowed grain in bulk (archaeologically attested in Roman times) without shovelling: this would have led to an incredibly high rate of wastage hardly compatible with the existence of surpluses needed to feed city-dwellers and soldiers (on the need for shovelling, similar view in Matterne 2001, 150; on wastage, see below). Shovelling is apparently not needed when grain is stored hulled or mixed with chaff (Beutler 1981, 36; in the second half of the eighteenth century, this storage method was apparently disappearing; see also Sigaut 1981, 165, and 1988, 13–14).

Given these various factors and the absence of ancient documentation on the matter, not to mention the fact that the situation may have varied with time and space, it is very hard to give an estimate of free space. Recently, Jean-Daniel Demarez, studying the granaries from Alle, put forward the hypothesis that storage in heaps would have left 70% of the granary unused (and "only" 40% if produce was stored in bins: Demarez et al. 2010, 393–94). However, this is nothing more than an educated guess and it seems safer to turn to historically attested examples. I found two, both from the eighteenth century and both for threshed and winnowed grain in bulk. Very recently (2016), Enrico Da Gai and Giulia Vertecchi published a 1788 document regarding the calculation of the storage capacities of the public granaries in Terra Nova in Venice (buildings now destroyed, but well documented). They arrive at a maximum usable surface area of ca. 85% of the total area. This is very close to the conclusions reached by Duhamel du Monceau in 1768 about the maximum usable surface area of the Grenier d'Abondance from Lyon. He calculated a maximum value of 81% of the total area, but noted himself that the available free space would be barely sufficient (Duhamel du Monceau 1753, 235–36, 1768, 249–51). Earlier in his book, making a theoretical calculation, he had come to the conclusion that 68.5% of the total area could be used for storing grain in heaps (Duhamel du Monceau 1753, 14–16, 1768, 13–16). Interestingly, the space needed for shovelling only accounts for less than 3% of the total area of the granary; but this is in stark contrast with the recommendations by Paul Diffloth in 1907, who wrote that to each heap of grain should correspond the same surface of free space, in order to shovel the grain from one spot to another (Diffloth 1907, 360).[7] All in all, it seems advisable to consider that when grain is stored in bulk, be it threshed and winnowed, with chaff or in spikelets, at least 1/3 of the granary needs to be kept free. Further research is needed to determine whether the examples quoted here are representative of ancient practices.

The fourth parameter is closely linked to the previous one and concerns the height of the grain heap. Figures found in the literature vary from 20 to 490 cm. In Table 3.2, I have compiled as much data on the subject as was available to me (but with no claim to exhaustivity). I have classified it according to the nature of the reasoning: unknown, based on the architecture of the storage building, based on agronomy and/or historical examples (it is often hard to distinguish between the two in older literature), based on archaeological evidence.

Table 3.2 makes it very clearly that a mean height of 20 to 40 cm is consistent with archaeological data, agronomic literature and historical cases from the seventeenth to the early twentieth centuries. This margin, given by almost every author for fresh grain (Reneaume is an exception), can be applied with confidence to the ancient countryside: since harvests were annual, grain stored on rural sites was mostly fresh. The discovery of Amiens "ZAC Cathédrale" shows that these values apply both to hulled and naked or dehusked grain. Since archaeological finds consist mainly of charred grain, it is important to note that charring has a significant impact on the weight but not on the volume of the grain (Ferrio et al. 2004, in particular 1636, tab. 1 and 1638, tab. 2). It should be stressed that layers might not always be fully preserved, although it appears to be the case in Amiens.

The maximum height of the heap for dry grain is much more variable. Sigaut gives 100 cm as a maximum for grain stored in bulk and shovelled; in the late eighteenth century, similar values are given by Krünitz and Tessier. But in

7 Various scholars, including Ferdière in this volume (Chapter 5), cite a later edition (1917) of Diffloth's work. I have not been able to consult it, but the contents appear to be the same.

the early twentieth century, Diffloth and the *Larousse agricole* give lower figures, as does the Venetian archival document from 1788 cited above. The finds from Ribchester (up to 60 cm) and Engers (several feet at some points), both military, may thus have consisted of dry grain.

Of course, the grain heap cannot be entirely flat. Grain being a semi-fluid, the heap is conical in shape. It is stable when the angle is within the range of 25 to 30° (De Lucia and Assenato 1992 [fr] = 1994 [en], appendix 2; Brinkkemper 1993, 150; see also Monteix in Boetto et al. 2016, 215 with note 113. In this volume, Ferdière opts for an angle of 45° which may thus be too steep). In theory, then, the wider the heap, the higher it may be. Otto Brinkkemper (1993, 149–50), for instance, trying to work out the storage capacity of two granaries with a usable surface area of ca. 16 m^2, reconstructs a single heap reaching 115 cm, with a 30° slope, equivalent to an even layer of grain 30 cm thick. Although Brinkkemper deals with hulled crops, this raises questions about the preservation of the grain in the centre of the heap. Several small heaps seem more likely: such is the case in the photographs from a modern granary in Heilbronn republished by Schubert (2016, 337). Furthermore, when different crops are stored in the same room, a case often encountered in excavations, they are always stored separately. In our calculation, assuming an even layer with a mean thickness of 30 cm should therefore be a good approximation of real conditions in ancient times.

In the light of the evidence presented in Table 3.2, figures above 100 cm for grain storage in bulk appear overestimated. Grain cannot be stored for any length of time in such conditions and too high a heap renders shovelling very impractical, if not impossible. If one accepts such values, one must also assume very high wastage. Estimating wastage during storage and calculating storage capacities are two different issues, but both must be tackled if we are to understand how storage facilities were managed and what proportion of agricultural surplus was actually available.

André Tchernia noted that a 20–25% wastage rate was generally assumed for granaries of the Roman period, but that the figure was not supported by scientific studies (Tchernia 2011, 245–46 = 2016, 194; see Papi and Martorella 2007, 90 for a 25% wastage rate given without justification). Véronique Matterne arrives at a similar figure for Amiens "ZAC Cathédrale", with 20% of the stock unfit for human consumption and 90% germinated (Matterne, Yvinec, and Gemehl 1998, 109–10. See also Smith and Kenward 2011 on pests, whose effects were overestimated by Buckland 1978). But we are dealing here with consumer, not production sites. Sigaut (1981, 165) wrote that on the whole peasants tend to take better care of their stocks and to reduce wastage; indeed, research on developing countries (mainly from the southern hemisphere) reports

TABLE 3.2 Height of grain heaps in the historical, agronomic and archaeological literature

Type source	Publication	Height cm
Unknown	Haverfield and Collingwood 1920	ca. 180 cm
	Richmond and McIntyre 1938–1939	ca. 150 cm
	Herzig 1946	300 cm
	Manning 1975	150 cm
	Galsterer 1990	125 cm
	Brinkkemper et al. 1995	30 cm
	Bakels 1996	100 cm
	Mauné and Paillet 2003	40 cm
	Demarez et al. 2010	30 to 100 cm
Architecture	Gentry 1976	up to 300 cm
	Papi and Martorella 2007	300 to 500 cm
Architecture/ agronomy	Hermansen 1982	240 to 400 cm
	Boetto et al. 2016	160 cm (max. 220 cm)
Agronomy/ history	Archives, 1684	ca. 93 cm
	Reneaume 1730 [1708]	ca. 60 to 80 cm
	Duhamel 1768	ca. 45 cm
	Duhamel 1768	ca. 30 cm
	Galiani 1770	60 to 75 cm
	Krünitz 1788	30 cm
	Krünitz 1788	ca. 60 cm
	Krünitz 1788	90 to 120 cm
	Parmentier 1789	30 to 45 cm
	Tessier 1793	ca. 45 cm
	Tessier 1793	ca. 105 cm
	Tessier 1793	ca. 150 cm
	Da Gai and Vertucchi 2016	ca. 34 to 68 cm
	Mathieu de Dombasle 1862	25 to 40 cm

Region or site discussed	Remarks
Great-Britain	Theoretical value. Storage in bins
Great-Britain	Theoretical value. Storage in bins
Windisch (CH)	Theoretical value. Storage in bins
Great-Britain	Theoretical value. Storage in bins or bags
	Maximum value based on contemporary practices
Netherlands	Mean value for a storage in a cone-shape heap attaining 115 cm in height at the top
Vareilles (FR)	
Alle (CH)	Values probably taken from agronomy treatises
Corbridge (GB)	Calculation based on wall thickness; bulk storage. Note that Gentry thought that grain was stored in sacks and not in bulk.
Thamusida (MA)	Calculation based on wall thickness; bulk storage
Ostia, Portus (IT)	Hermansen relies on parallels from present-day Canada and adapts the value according to the height of the rooms.
Ostia (IT)	The authors do no think that heaps reached 220 cm, which is given as a maximum value.
Abtei Weingarten (DE)	Maximum height. Hauptstaatsarchiv Stuttgart B 522 Bü 69. (reference supplied by Lars Blöck)
France	For fresh (wet) grain; with time grain settles and the height decreases
France	Mean values observed in public granaries
France	Very wet grain, 1745 harvest
France ?	2,5 feet = maximal height
Germany	For wet grain
Germany	For dry grain
Germany	Maximum height for dry grain
France	
France	Normal mean height
France	1-year-old grain
France	2-year old grain or older
Venice, Rome (IT)	For bulk storage, depending on dryness. 18th c. archives
France	Posthumous, author died in 1843

TABLE 3.2 Height of grain heaps in the historical, agronomic and archaeological literature (cont.)

Type source	Publication	Height cm
Agronomy/history (cont.)	Heuzé 1876	60 cm
	Desmoulins 1896	25 to 30 cm
	Desmoulins 1896	50 to 60 cm
	Diffloth 1907	15 to 40 cm
	Diffloth 1907	up to 60 cm
	Ernest and Dumont 1921	80 cm
	Sigaut 1981	100 m
	Malrain 2000	40 cm
	Geraci 2015 = Geraci and Marin 2016	30 and 70 cm
Archaeology	Knörzer 1970	10 to 20 cm
	Gentry 1976	10 to 60 cm
	Matterne, Yvinec, and Gemehl 1998	20 to 40 cm
	Bouby 2001	3 to 15 cm in one spot; 35 cm in another spot
	Dorow 1826	9 to 24 cm, but several feet in places
	Matterne 1997	20 to 40 cm
	Ruas et al. 2005	5 cm
	Ruas 2003	up to 10 cm
	Borgongino 2006	20 cm

Region or site discussed	Remarks
France	Seems to be a maximum height
France	For grain just harvested
France	Mean height, when grain has dryied
France	For fresh grain, depending on dryness. Matterne et al. 1998, 107 cites 4th ed. (1917) with same numbers
France	For very dry grain in well-built granaries; mean values are lower (gives 33 cm as an example)
France	Maximum height for bulk storage with shoveling
General	Maximum height for bulk storage with shoveling
Northern France	Discussion with agronomist
General	Maximum heights for fresh and dry grain. Figures taken from mid-20th c. Italian agronomy treatises
Neuss (DE)	1st c. CE. Naked wheat
Ribchester (GB)	Military (northern granary)
Amiens (FR)	2nd c. CE, urban context. Mainly spelt
Crest "Bourbousson" (FR)	3rd c. CE, rural context. Mainly naked wheat, barley in one spot (publication does not specify if hulled or dehusked)
Engers (DE)	4th c. CE, military post. Mainly naked wheat, with rye and barley
Compiègne (FR)	10th c. CE. Mainly rye and naked wheat
L'Isle-Jourdain "La Gravette" (FR)	11th c. CE. Mainly naked wheat, hulled grain stored in chaff
Durfort "le Castlar" (FR)	14th c. CE. Mainly rye and millet; hulled grain at least partially stored in spikelets
Herculaneum "Villa dei Papiri" (IT)	1st c. CE. Barley

TABLE 3.3 A tentative model for calculating grain storage capacities

Points to be considered	Propositions
Type of cereal stored	See Ouzoulias 2006, 173–177 and other relevant literature, for the various densities
Hulled or husked grain	See Ouzoulias 2006, 173–177 and other relevant literature, for the various densities
Usable surface area	Mean value of 70 % of total floor area? To be investigated further
Thickness of layer	20 to 40 cm (mean value of 30 cm)
Loss during harvest and storage	Mean value of 10 %? To be investigated further

post-harvest losses of 10 to 15%, part of which occurs after rural storage (*Post-harvest Food Losses in Developing Countries* 1978; De Lucia and Assenato 1992). Moreover, grain considered wasted by current standards might still have been consumed or used in other ways in Antiquity (Matterne, Yvinec, and Gemehl 1998, 110; Smith and Kenward 2012, 146–48). We are a long way from the 50 to 80% wastage rate considered possible by Thomas Gallant (1991, 97–98) in his study of ancient Greece, which seems completely implausible to me,[8] as well as hardly compatible with the low yields he estimates (yields are discussed at more length in Garnsey, Gallant, and Rathbone 1984. On this subject, see also Halstead 2014, 238–50. More generally, on the productivity of ancient peasants, see Kron 2008).

Table 3.3 is an attempt to sum up the different topics discussed in this chapter, and to provide guidelines for calculating the storage capacities of rural granaries. This "model" is put to test in Chapter 6 of this volume, where the reader will find a discussion of historical implications of storage capacity estimates for the Rhine-delta area.

As a general rule, it appears that so far, estimates put forward by scholars for urban and military *horrea* are much too high. They should certainly not be transposed to rural granaries. Conversely, it must be stressed again that the tentative model presented here is meant for rural sites from the North-Western provinces, and that the focus is on production sites, where fresh grain would enter the granaries each year. This would not necessarily have been the case in

8 I have not been able to consult the literature cited by Gallant but more recent works on post-harvest losses in developing countries do not come near a 50–80 % wastage rate during storage: see e.g. Aulakh and Regmi 2013; 'A Review of Methods for Estimating Grain Post-Harvest Losses' 2015, with previous literature.

urban and military contexts: the wheat found on the Woerden wreck had been stored for some time previous to its shipping. It should have been dryer than freshly harvested grain and thus presumably storable in higher heaps. Heaps higher than 30 to 40cm may also have been more common in granaries used for very short storage periods during transhipments: such may have been the case for the *horrea* in Portus.

To conclude, the model presented here should be critically assessed before being used or adapted to other types of sites or to another region. This is also true of the historical and ethnographic *comparanda* we use to compensate the silence of ancient sources. I have been careful to use here mainly French material referring to pre- or early industrial agriculture: although it appears consistent with similar material from Germany and Italy, its general validity for the Western Roman Empire still has to be assessed.

PART 2

Storage through Time and Space

CHAPTER 4

Évolution du stockage agricole dans la moitié septentrionale de la France à l'âge du Fer (VIe–Ier s. av. n. è.)

Stanislas Bossard

À l'aube du XXIe s., la thèse de doctorat de Frédéric Gransar a posé les principaux jalons de l'évolution du stockage alimentaire en Europe tempérée au cours de l'âge du Fer, et reste aujourd'hui une étude de référence (Gransar 2001). En fondant son propos sur l'analyse des structures de stockage de 346 sites répartis sur douze pays, l'auteur examine les aspects techniques, fonctionnels et socio-économiques de la conservation des denrées agricoles depuis leur production jusqu'à leur consommation. Il démontre ainsi que le stockage alimentaire connaît d'importantes variations quantitatives et qualitatives au cours des siècles, témoins des changements organisationnels des populations de l'âge du Fer d'Europe tempérée. Les données qu'il exploite sont issues d'opérations archéologiques essentiellement récentes, notamment pour le Nord de la France, secteur auquel il accorde une attention particulière (pour ce secteur, voir aussi Gransar, Matterne, et Pommepuy 1999; Gransar 2000, 2003). De fait, pour le territoire français, le développement considérable de l'archéologie de sauvetage puis préventive a multiplié les exemples bien documentés d'établissements ruraux gaulois à partir des années 1980.

Depuis les travaux fondateurs de F. Gransar, en ce qui concerne le territoire métropolitain, les recherches menées sur ce sujet se sont attachées à des secteurs plus restreints, à l'occasion de quelques synthèses régionales. Des enquêtes récentes apportent un nouvel éclairage sur deux zones peu étudiées dans la thèse de F. Gransar : d'une part l'Est de la France, en particulier les vallées de la Moselle, de la Meurthe et du Rhin entre le XIe et le IIIe s. av. n. è. (Deffressigne et Landolt 2017), et d'autre part, à plus de 500 km à l'ouest, la péninsule armoricaine et sa périphérie[1].

1 Ce secteur a bénéficié, outre d'un bilan portant sur les différentes architectures de stockage employées du VIe au Ier s. av. n. è. (Bossard et al. 2018), de deux synthèses consacrées à des types de structure en particulier, les souterrains et caves boisées en Bretagne et Normandie occidentale ainsi que les greniers surélevés de La Tène moyenne au début de l'époque romain en Pays de la Loire.

DOI:10.1163/9789004389045_006

La base de données "datAFer2" dédiée à l'habitat rural de l'âge du Fer, mise en place sous l'égide de l'Inrap en 2005 et depuis alimentée régulièrement[2], fournit un outil pratique et efficace pour réfléchir à la question du stockage alimentaire à une large échelle. En effet, elle précise, pour chaque état des établissements intégrés à la base, le nombre et le(s) type(s) de structure de stockage mis en évidence lors de la fouille ou du diagnostic archéologique. En élargissant le cadre géographique adopté lors d'études récentes à la moitié septentrionale de la France (Zech-Matterne, Auxiette, et Malrain 2013; Zech-Matterne et al. à paraître), cet article propose de définir les principales évolutions du stockage alimentaire à la fin de la Protohistoire, en s'appuyant sur les données renseignées au sein de cette base, tout en tenant compte des compléments apportés par les synthèses régionales réalisées depuis les travaux de F. Gransar.

1 Cadres et limites de l'étude

La chronologie considérée pour cette enquête, du VIe s. à la fin du Ier s. av. n. è., couvre la fin du premier âge du Fer (Hallstatt D) et le second âge du Fer, jusqu'à La Tène D2b. L'emprise géographique choisie concerne 13 régions renseignées dans la base de données, correspondant globalement à la moitié nord de la France[3]. Notons d'emblée que les résultats présentés ici restent largement tributaires de l'archéologie préventive, impliquant une répartition géographique des sites fouillés très inégale (fig. 4.1). Ainsi, certains tracés linéaires, liés à des travaux d'aménagement routier ou autoroutier, sont privilégiés, de même que la périphérie de plusieurs métropoles françaises (Rennes, Caen, Arras ou Compiègne) a bénéficié d'un développement industriel ou urbain favorisant la découverte de vestiges d'époque gauloise en contexte préventif. S'ajoutent à ces secteurs bien documentés la vallée de l'Aisne, à l'est de Soissons, ainsi que la plaine d'interfluve Seine-Yonne, où l'exploitation de carrières a conduit à la mise au jour de nombreux établissements ruraux, ou encore la vallée de la Moselle et la Plaine de France, au nord de Paris. En revanche, des zones de vide ponctuent le territoire étudié : la Franche-Comté, la Bourgogne, le

2 Base coordonnée par Geertrui Blancquaert, François Malrain et Thierry Lorho, disponible en ligne à l'adresse : <http://agedufer.inrap.fr>. Pour une présentation de cet outil, voir Blancquaert et al. 2009 et Malrain, Blancquaert, et Lorho 2013.

3 Le territoire pris en compte intègre les régions françaises telles qu'elles étaient définies avant la réforme territoriale opérée en janvier 2016 : Bretagne, Pays de la Loire, Basse et Haute-Normandie, Centre, Île-de-France, Picardie, Nord-Pas-de-Calais, Champagne-Ardenne, Bourgogne, Lorraine, Franche-Comté, Alsace.

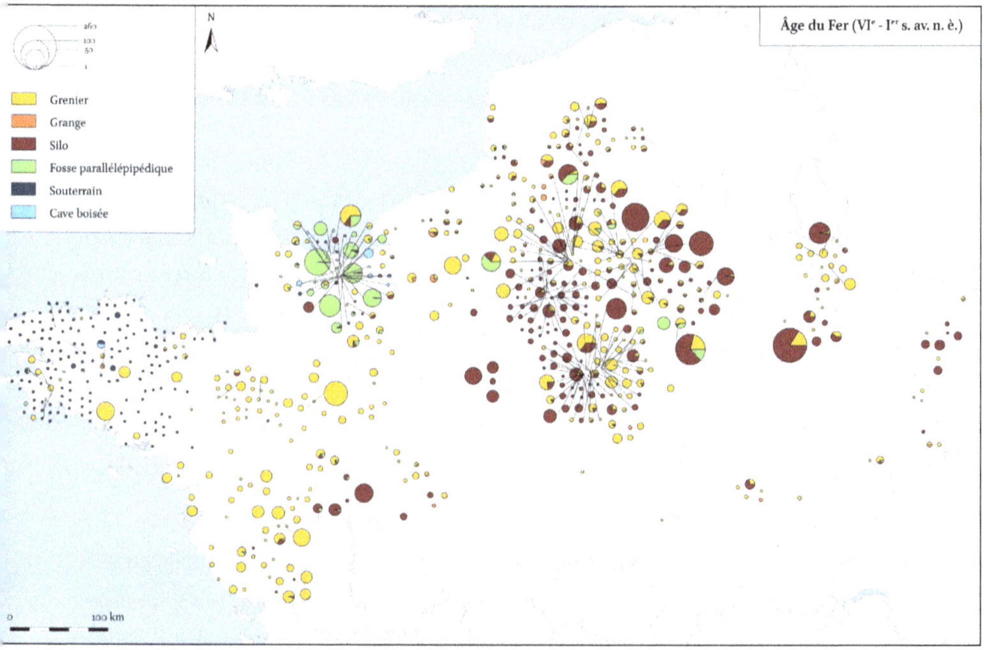

FIGURE 4.1 Répartition géographique des structures de stockage identifiées du VIe au Ier s. av. n. è.
S. BOSSARD

Nord-Pas-de-Calais et le Centre de la France n'interviendront que ponctuellement au cours de cette réflexion.

La documentation utilisée se répartit en deux types de sources : pour l'Ouest de la France, les données retenues sont celles collectées dans le cadre d'une synthèse entreprise en 2016 (Bossard et al. 2018), et sont donc issues du dépouillement systématique de l'ensemble des rapports d'opérations archéologiques menées sur les établissements gaulois de trois régions (Bretagne, Pays de la Loire et Basse-Normandie). Pour le reste du territoire ici considéré, seules les informations chiffrées enregistrées au sein de la base de données des habitats ruraux gaulois ont été prises en compte[4]. L'alimentation de la base se faisant à des rythmes variés en fonction des régions, la mise à jour des données n'est pas uniforme à l'échelle de la moitié septentrionale de la France ; certains secteurs n'intègrent pas les découvertes les plus récentes. En outre, l'identification d'une structure de stockage peut être équivoque dans le cas d'un grenier aérien ou d'une grange dont il ne subsiste que les trous de poteaux, ou encore

4 Nombre de greniers, silos ou fosses parallélépipédiques par phase d'un établissement rural.

d'un silo enterré au profil mal conservé : d'une région à l'autre, le nombre d'architectures caractérisées dans la base de données en tant que structures de stockage peut donc fluctuer suivant l'appréciation de l'archéologue qui élabore les notices de site.

En tenant compte de ces biais méthodologiques induisant quelques imprécisions, voire des erreurs ponctuelles – corrigées ici lorsqu'il a été possible –, 4623 structures de stockage réparties sur 583 sites ont été retenues. Les cartes de répartition réalisées par siècle (voir *infra*) se limitent aux structures chronologiquement bien calées ; celles dont la datation proposée couvre plus de deux siècles ne figurent alors pas sur ces cartes.

2 Variété et spécificités régionales des structures de stockage

Au cours de l'âge du Fer, les Gaulois ont construit ou creusé au sein de leurs établissements ruraux différents types de structures qu'ils ont dédiées au stockage de leurs productions agropastorales, pour une durée variant de quelques jours (dans le cas d'une consommation rapide des aliments) jusqu'à plusieurs mois voire années, notamment pour les récoltes destinées à être ensemencées ou à être échangées ultérieurement. Tandis que les traditionnels greniers surélevés sur quatre, cinq, six ou neuf poteaux (fig. 4.2) ont investi l'ensemble du territoire considéré, les autres catégories de structures ont été préférentiellement utilisées sur des aires géographiques spécifiques (fig. 4.1). Ainsi, des silos ont été reconnus dans le quart nord-est de la France et dans la vallée de la Loire (fig. 4.2), et des souterrains et caves boisées ont été employés dans l'ouest de la péninsule armoricaine, mais aussi dans la Manche et le Calvados où ils sont généralement associés à des fosses parallélépipédiques (fig. 4.3 et 4.4). Les granges identifiées au sein de la base de données sont à considérer avec précaution : ces grands bâtiments sur poteaux porteurs peuvent avoir revêtu des fonctions variées, difficiles à déterminer aujourd'hui (écuries, annexes diverses). Enfin, au sein des habitations, de grands vases en terre cuite (enterrés ou non) ont aussi joué un rôle essentiel dans la conservation de petites quantités de produits consommés quotidiennement (Gransar 2000, 2001, 306–12; Deffressigne et Landolt 2017. Les contenants en terre cuite n'ont pas été intégrés à cette enquête, faute de données disponibles à l'échelle de la zone étudiée).

Les structures de stockage possèdent des avantages et des inconvénients propres à chaque type (pour une présentation détaillée, voir Gransar 2000; Bossard et al. 2018). Par exemple, le silo souterrain, rapide et peu coûteux à aménager, permet la conservation de grandes quantités de denrées végétales

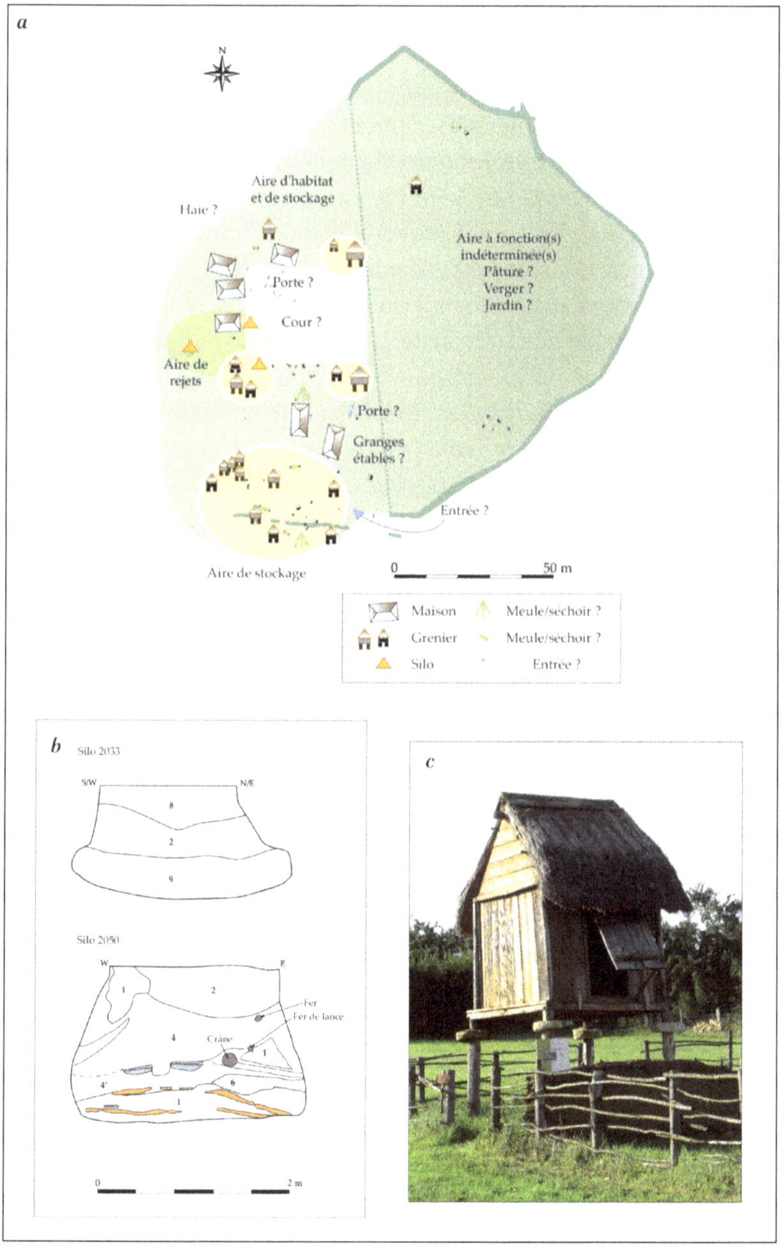

FIGURE 4.2 a) Greniers de l'âge du Fer au sein d'un établissement rural fouillé à Poulainville (Somme), où ils sont associés à des silos souterrains (d'après Malrain et al. 2016) ; b) coupes de deux silos de Poulainville (ibid.) ; c) reconstitution d'un grenier aérien au musée archéologique de Vieux-la-Romaine (cliché H. Lepaumier, Inrap)

sur le long terme. Son principe de conservation reposant sur l'anaérobie, il n'est guère adapté à un accès répétitif, mais se révèle utile pour entreposer de grandes quantités destinées à être consommées en une seule fois, telles que les semences. En revanche, le grenier aérien offre un accès simple et répétitif aux produits stockés, convient à tous les types de milieu (sec à humide), mais nécessite de brasser régulièrement les grains entreposés pour empêcher tout risque de germination ou de combustion spontanée induite par les dégagements gazeux. Si l'on ne connaît pas les denrées qui étaient entreposées à plus de deux mètres sous terre dans les souterrains ou les caves boisées de l'Ouest de la Gaule, il est probable que l'obscurité, la fraîcheur et l'humidité de ces volumes enterrés aient joué un rôle prépondérant dans les modalités du stockage. Il s'agit néanmoins de vastes structures qui ont requis des efforts considérables pour être creusées : constituées d'une à plusieurs salles reliées par des passages étroits ou des couloirs, elles communiquent avec la surface à l'aide de puits et d'escaliers, l'ensemble pouvant se développer sur plusieurs dizaines de mètres.

Quels produits ont pu être conservés au sein de ces différentes structures ? Si les silos et peut-être les fosses parallélépipédiques semblent avoir été dédiés exclusivement au stockage de produits végétaux (céréales, légumineuses), les greniers, granges, souterrains et caves boisées ont pu aussi recevoir d'autres denrées, telles que de la viande, du poisson, des liquides alcoolisés ou non, des produits laitiers, voire des artefacts, tels que des outils. Pour ces quatre dernières catégories de structure, les modalités du stockage devaient être variées et peut-être complémentaires : en vrac, en sacs ou en paniers, voire en récipient en bois ou terre cuite. Rares sont les restes d'aliments mis au jour en association avec ces structures de stockage : pour les greniers par exemple, quelques études carpologiques ont été mises en œuvre à partir de graines de céréales découvertes dans les trous de poteaux ou à proximité du bâtiment (par ex. à Jaux dans l'Oise : Malrain et al. 1996; à Val-de-Reuil dans l'Eure : Moreau, Granier, et Dubreucq 2015; et à Entrammes en Mayenne : Guillier et al. 2015).

Quant aux raisons qui ont conduit les Gaulois à multiplier les types d'architectures de stockage, elles sont sans doute diverses. Bien que la géologie et la pédologie puissent expliquer l'absence de structures souterraines sur certains sols ou substrats, elles ne justifient pas à elles seules la diversité des types d'ouvrage connus. La nature des produits stockés, qui reste inconnue dans la plupart des cas, a certainement impliqué d'employer un type d'ouvrage adéquat aux conditions de conservation requises. Il semble pertinent d'admettre des usages complémentaires entre certaines structures, en raison de leurs caractéristiques propres évoquées ci-dessus. Les souterrains et caves boisées,

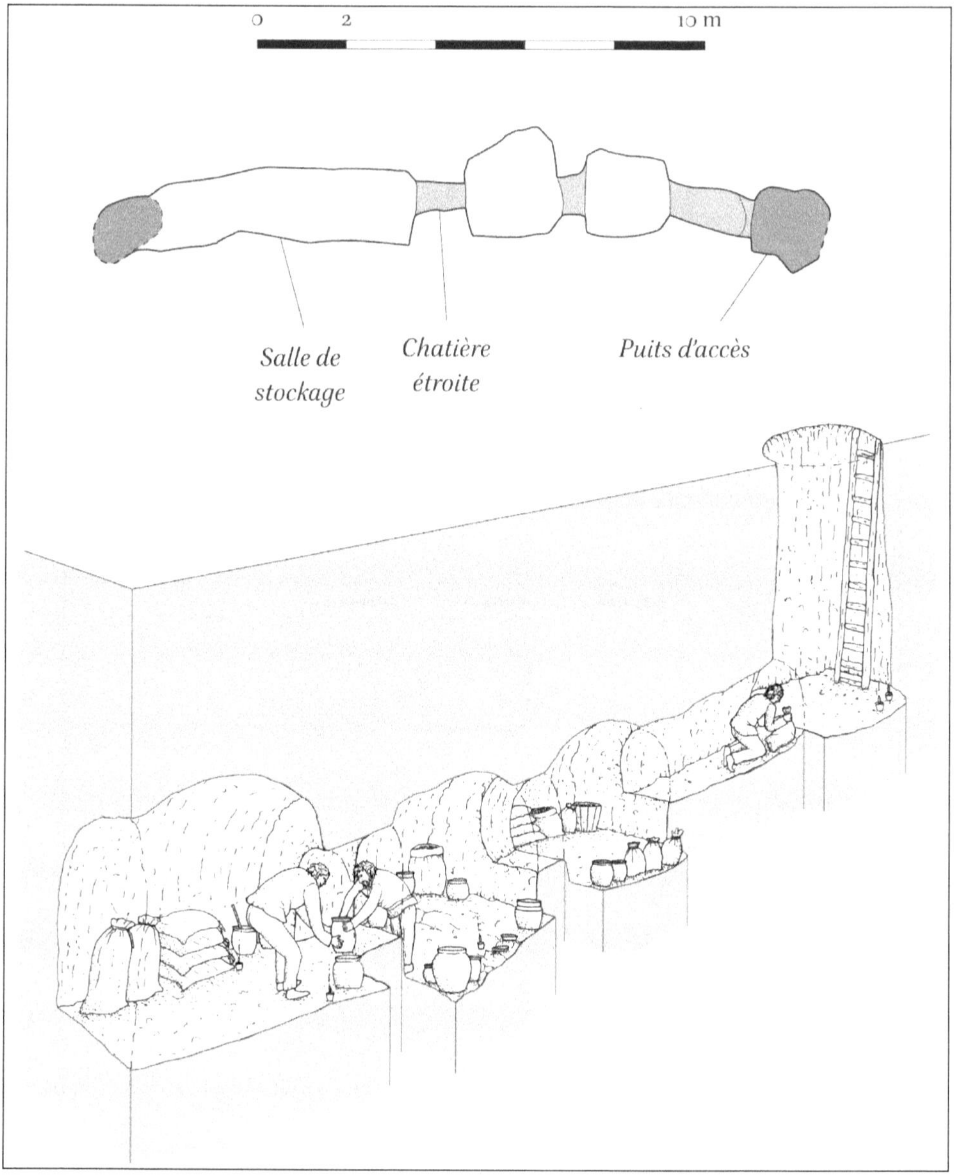

FIGURE 4.3 Plan et évocation graphique du souterrain de Plabennec (Finistère). Ces architectures sont composées d'un ensemble de salles destinées au stockage de denrées et reliées par des passages étroits (chatières).
S. BOSSARD

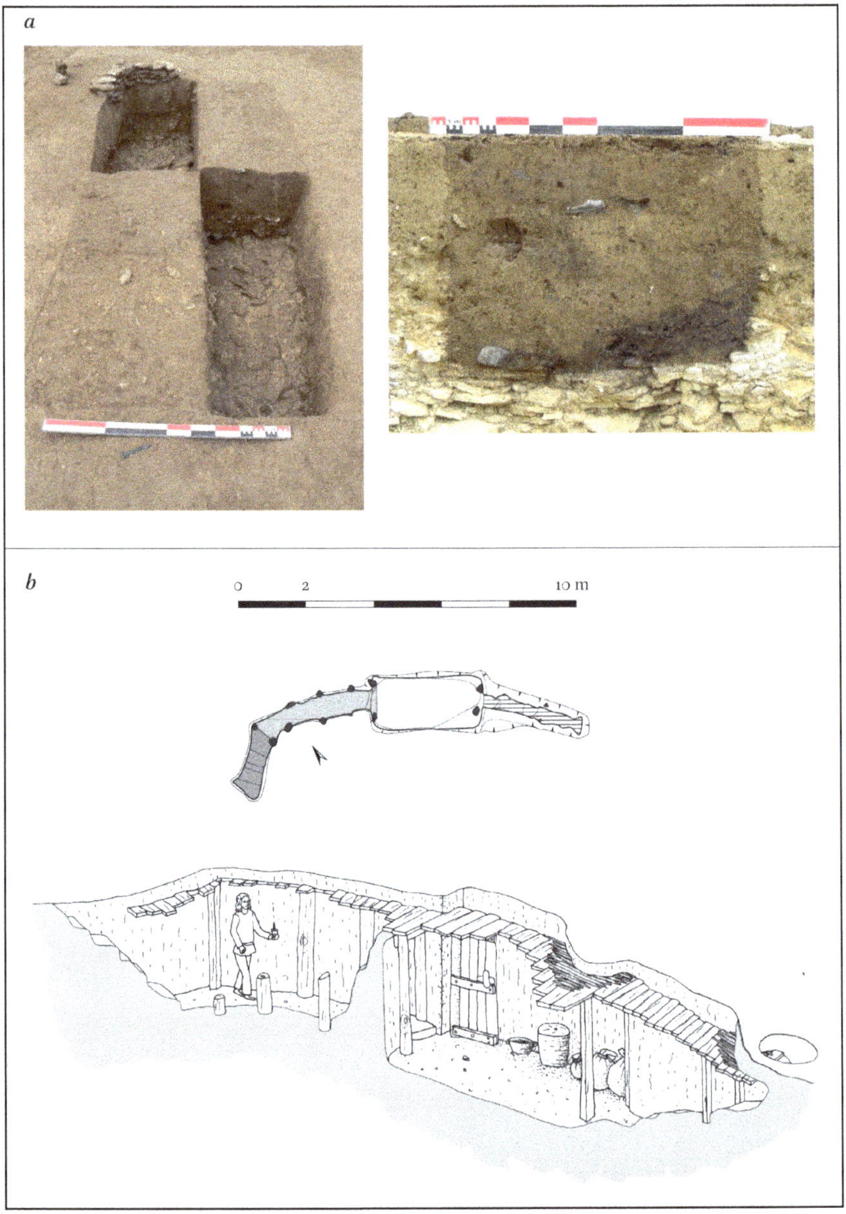

FIGURE 4.4 a) Fosses parallélépipédiques en cours de fouille et vue en coupe sur deux habitats gaulois de Fleury-sur-Orne (Calvados ; d'après Lepaumier 2012) ; b) plan et évocation graphique d'une cave boisée fouillée à Ifs (Calvados ; S. Bossard)

étroitement liés aux habitations dans l'Ouest de la Gaule (Bossard 2015, 161–81), ont manifestement assuré un accès régulier aux produits consommés dans les unités domestiques, à l'instar des céramiques dédiées au stockage, tandis que des silos ou greniers isolés ou regroupés à l'orée des champs recevaient vraisemblablement les graines à semer ultérieurement. En outre, les structures souterraines – silos, caves boisées, fosses parallélépipédiques et souterrains – ont pu servir de cachette pour les réserves alimentaires, entre autres en période de trouble. En tout état de cause, la répartition géographique des structures témoigne de phénomènes de modes culturelles localisées, qui sont en outre changeantes au cours de la période considérée. De fait, l'emploi préférentiel de certains types de structure de stockage est aussi fonction du temps, comme l'atteste l'évolution de ces aménagements tout au long de l'âge du Fer.

3 Évolution du stockage dans les campagnes de l'âge du Fer

Par commodité, un palier de cent ans a été choisi pour mesurer l'évolution des structures de stockage au cours de l'âge du Fer à l'aide de cartes de répartition des différents types d'architectures (*infra*, fig. 4.5 à 4.10). Ces cartes permettent d'apprécier les oscillations quantitatives et qualitatives qu'ont connues les aménagements liées au stockage alimentaire de la fin du Hallstatt à la conquête romaine, ainsi que les spécificités régionales. Deux grandes phases d'environ trois siècles chacune peuvent alors être distinguées, mettant notamment l'accent sur les profonds bouleversements qui ont affecté les campagnes gauloises dans le courant du IIIe s. av. n. è.

3.1 *Du Hallstatt final à la Tène moyenne* (*VIe s.– milieu du IIIe s. av. n. è.*)

D'une manière générale, le monde rural connaît lors de la seconde moitié du VIe s. av. n. è. une intensification de l'occupation du territoire, qui se couvre alors d'établissements ruraux ouverts ou enclos, au sein ou à proximité desquels on exploite une grande diversité d'espèces cultivées, céréales, légumineuses et oléagineux. La pratique de la méture, visant à la mixité des semis de manière à garantir la récolte, est attestée (Malrain, Blancquaert, et Lorho 2013, 225 et 230; Blancquaert et al. 2012). Du point de vue du stockage, les structures documentées du VIe au IVe s. av. n. è. sont essentiellement concentrées dans le sud de la Picardie et à la confluence de la Seine et de l'Yonne, en Seine-et-Marne ; d'autres découvertes apportent des éclairages ponctuels, mais les données demeurent lacunaires pour de nombreuses zones (fig. 4.5 à 4.8).

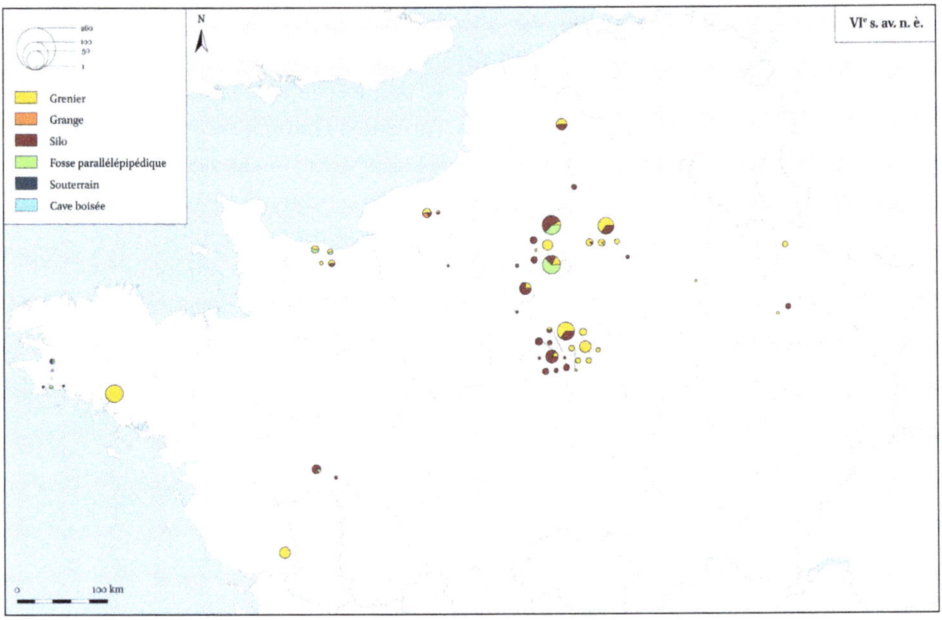

FIGURE 4.5 Répartition géographique des structures de stockage identifiées au VIe s. av. n. è.
S. BOSSARD

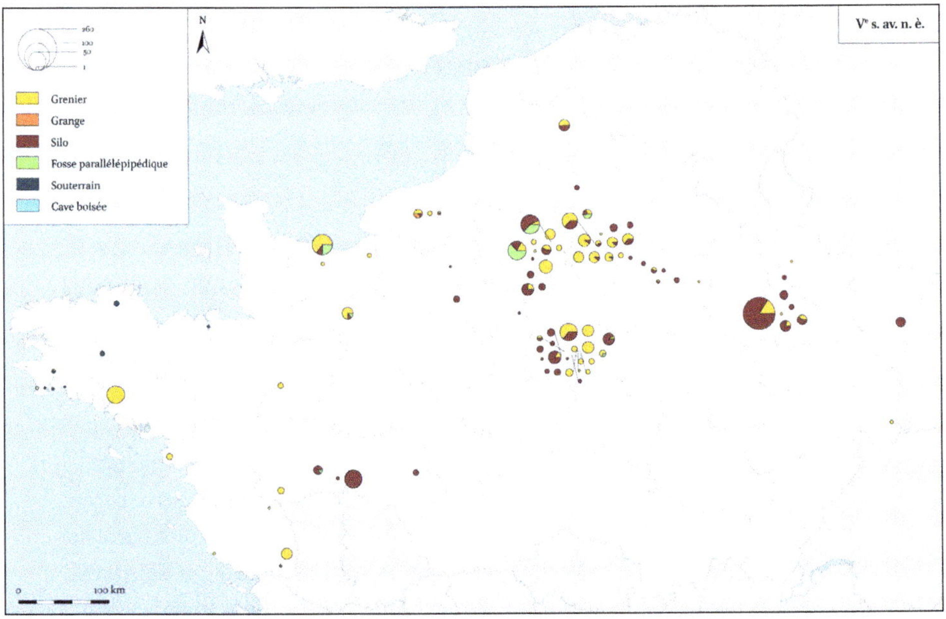

FIGURE 4.6 Répartition géographique des structures de stockage identifiées au Ve s. av. n. è.
S. BOSSARD

Pour cette première grande phase, à l'exception de la partie occidentale du Massif armoricain, les silos souterrains et les greniers aériens constituent les deux grands types d'architectures aménagées au sein des exploitations rurales pour y conserver les produits alimentaires. Sur la majorité des établissements, on les retrouve en nombre restreint, généralement inférieur à cinq, voire dix exemplaires. Ainsi, un à deux greniers sont associés aux unités domestiques des Étomelles à Villeneuve-Saint-Germain (Aisne) au VIe s., où ils devaient assurer la conservation des réserves nécessaires aux habitants, tandis que quatre greniers isolés à plus de 300 m de l'occupation ont pu contenir les semences, en bordure des parcelles cultivées (Auxiette et al. 2003). La même hypothèse a été émise pour la dizaine de silos qui se partagent entre habitat et périphérie sur la “dernière tranche” d'Ifs (Calvados) (Besnard-Vauterin 2011, 194).

Si les greniers et les silos semblent utilisés en proportions équilibrées au VIe et au début du Ve s., les seconds connaissent un succès certain à partir du Ve et surtout du IVe s. au milieu du IIIe s. Ils se multiplient alors dans des secteurs où ils étaient déjà présents, de manière plus discrète, dès le VIe s. : dans le bassin parisien, en Lorraine ou encore dans la vallée de la Loire, à hauteur de la Touraine et de la partie orientale du Maine-et-Loire.

Les fosses parallélépipédiques, creusements rectangulaires aux parois verticales et profonds de plusieurs décimètres à plus d'un mètre, sont employées dans l'aire étudiée dès le VIe s. Elles apparaissent alors systématiquement en association avec des greniers ou des silos, suivant des proportions variables. Quoi qu'il en soit, elles demeurent une solution marginale. La fonction de stockage n'est pas avérée pour l'ensemble d'entre elles, mais la plupart peuvent selon toute vraisemblance être considérées comme des caves de morphologie assez simple (Bossard et al. 2018). Ces excavations se rencontrent dans le Calvados, où elles deviendront lors de la phase suivante une composante essentielle des habitats ruraux, ainsi qu'en Maine-et-Loire et autour de Compiègne, dans l'Oise.

À l'extrémité occidentale de la Gaule, l'ouest de la péninsule armoricaine constitue un cas à part, dès le VIe s., et ce jusqu'au IIIe ou IIe s. av. n. è. Ici, point de silos ni de fosses parallélépipédiques, et les greniers sont globalement absents. En revanche, quasiment chaque établissement semble doté d'un à plusieurs souterrains voire, pour les VIe et Ve s., de caves boisées (Bossard 2015). Ces architectures enterrées apparaissent également plus à l'est, dans le Calvados et la Manche, au plus tard à la fin du Ve s. ou au début du IVe s. ; dans tous les cas, elles sont creusées le plus souvent à l'aplomb des maisons ou à proximité immédiate. En l'absence d'autres structures de stockage identifiées sur la péninsule armoricaine lors de cette phase ancienne, doit-on considérer que les souterrains servaient à la fois de garde-manger accessibles depuis l'habitation et de réserves invisibles depuis la surface, où stocker avec sûreté le produit des récoltes ? Faute de données carpologiques ou d'approche expérimentale, on ne sait si les céréales ou d'autres végétaux pouvaient être conservés, à moyen ou long terme, au sein de ces salles plongées dans l'obscurité et humides. Comme l'avait signalé F. Gransar, l'Ouest, en ce qui concerne le stockage, se distingue nettement des autres régions de la France et s'inscrit dans une zone qu'il qualifie d'occidentale et à laquelle il rattache également l'ouest et le nord des îles britanniques (Gransar 2001, 482–89). Si la réalité et l'homogénéité d'une telle zone occidentale peut être discutée, notamment pour la phase suivante, il peut être néanmoins noté que des constructions souterraines en pierre sont également attestées à la pointe occidentale de l'Angleterre, en Cornouailles, dès La Tène ancienne (Bossard 2015, 212–15). L'Ouest de la Gaule et la Cornouailles adoptent donc un mode de stockage similaire, témoignant de relations transmanches et d'échanges d'idées, et reflétant peut-être des pratiques agropastorales semblables.

Du VIe au IIIe s., certains sites ruraux se dotent d'une quantité considérable de structures de stockage, traduisant vraisemblablement une centralisation des récoltes au sein d'habitats ou de structures parfois isolées de toute occupation. Ce phénomène, mis en évidence par F. Gransar dans la "zone moyenne" de l'Europe (Gransar 2001, 336–43 et 499–506) – du Centre de la France à l'Allemagne, et des Alpes au sud-est du Royaume-Uni –, trouve également plusieurs échos dans l'Ouest de la France.

Des batteries de greniers sont alors installées, aux VIe et Ve s., au sein ou aux abords de résidences encloses que l'on peut qualifier d'aristocratiques grâce à plusieurs critères, tels que l'emprise au sol de l'habitat, la présence d'aménagements monumentaux (clôture, porche d'entrée, habitations) ou encore de mobiliers particuliers. En 2001, F. Gransar cite ainsi, entre autres, l'établissement des Terres du Bois Mortier à Grisy-sur-Seine (Seine-et-Marne), système d'enclos englobant deux grands bâtiments dont la monumentalité était probablement ostentatoire. À une centaine de mètres a été observée une batterie d'au moins 15 greniers implantés en arc de cercle. Il évoque au même titre la concentration de 17 greniers rassemblés dans un enclos palissadé au Grand Marais (Bucy-le-Long, Aisne) (Gransar 2001, 339). L'habitat partiellement fouillé de la Plaine du Bosc Renault à Hautot-le-Vatois (Seine-Maritime) offre un autre exemple de ce type de site (Blancquaert et Adrian 2006). Une douzaine de greniers, *a minima*, prend place au sein d'un enclos fossoyé dont l'habitation est une construction à abside vaste de 110 m². Plus à l'ouest, dans le Calvados, l'établissement de la Fosse Touzé à Courseulles-sur-Mer, daté du Ve et de la première

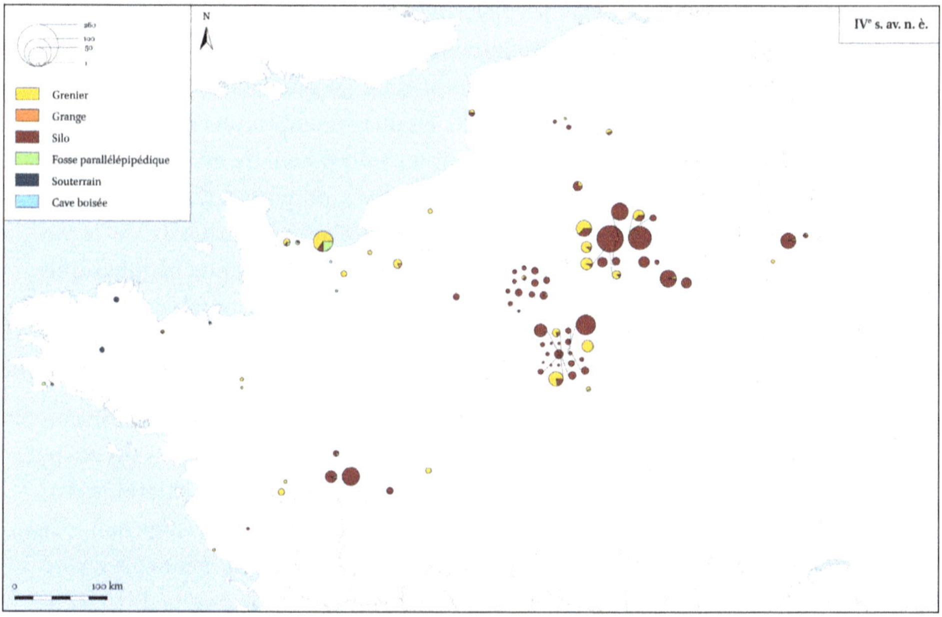

FIGURE 4.7 Répartition géographique des structures de stockage identifiées au IVe s. av. n. è.
S. BOSSARD

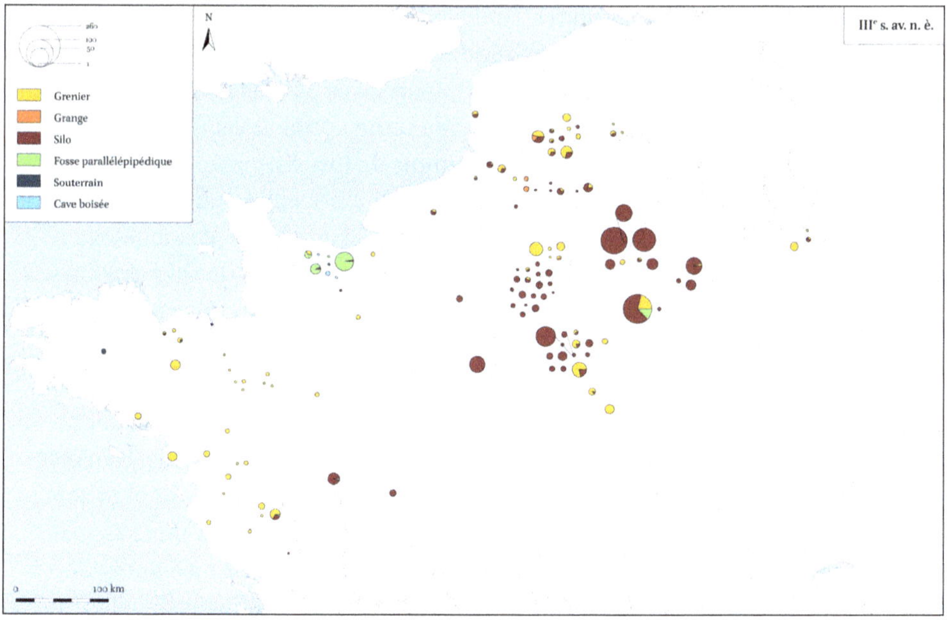

FIGURE 4.8 Répartition géographique des structures de stockage identifiées au IIIe s. av. n. è.
S. BOSSARD

moitié du IVe s., a été qualifié de “place de collecte monumentale” par I. Jahier (2011, 178–82). Il se compose, autour d'une place centrale, d'un grand bâtiment de plan circulaire et d'une batterie de plus de 30 greniers, l'ensemble étant protégé par une enceinte monumentale. Dernier exemple, un rôle de stockage alimentaire peut être aussi envisagé pour les trois ou quatre grands bâtiments dotés de quatre nefs délimitées par des rangées de poteaux, concentrés au sud du plateau Saint-Marcel, au sein du complexe aristocratique de Vix (Côte-d'Or). Ces édifices, mis en évidence au cours de prospections géophysiques menées dans les années 2000, sont considérés comme de probables entrepôts en raison de leur plan qui évoque les *horrea* d'époque romaine (Chaume et al. 2011, 376–78).

Ces concentrations de bâtiments sur poteaux porteurs, placés sous le contrôle direct des élites, offrent une capacité de stockage qui dépasse sans aucun doute les besoins de l'habitat. Assurent-ils la collecte des productions d'un vaste domaine, ou bien correspondent-ils à des espaces de stockage collectifs placés sous la protection de l'aristocratie ? Une autre question concerne la redistribution de ces produits. Ils peuvent être réunis dans l'optique d'être commercialisés à une échelle régionale ou suprarégionale ; selon F. Gransar, ces denrées n'étaient pas exportées vers les villes méditerranéennes, mais intervenaient plutôt dans le cadre d'échanges internes à l'Europe tempérée, à une échelle qu'on ne peut déterminer (Gransar 2001, 358–60). Outre les batteries de greniers clairement associées aux résidences rurales des élites gauloises, notons que d'autres concentrations de bâtiments de ce type ont été mises en évidence au sein de sites *a priori* dépourvus de statut social élevé. Tel est le cas de l'habitat de la Bouche à Vesle sur la commune de Ciry-Salsogne, dans la vallée de l'Aisne, doté d'une vingtaine de greniers regroupés en périphérie de l'habitat (Gransar 2001, 339; Hénon et al. 2002). Le rôle d'un tel type d'établissement agricole, produisant des surplus ou centralisant les récoltes de sites satellites, est difficile à définir, d'autant plus qu'il n'est guère aisé d'évaluer les besoins réels des habitats, *a fortiori* lorsque leur fouille n'a été que partielle.

Dès le Ve s., mais surtout au IVe s. et lors de la première moitié du IIIe s., les batteries de greniers surélevés disparaissent et laissent place à des concentrations de silos identifiées dans plusieurs régions. Contrairement aux greniers, les regroupements de silos sont *a priori* isolés dans le paysage, ou bien sont associés à des vestiges d'habitat modeste qui, dans tous les cas, ne se rapportent pas à une demeure aristocratique. Les fouilles réalisées à Gondreville – Fontenoy-sur-Moselle (Meurthe-et-Moselle) offrent des éléments de réflexion intéressants (Deffressigne et al. 2002). Une batterie de 131 silos dont les recoupements sont rares a été datée du Hallstatt D3, soit du début du Ve s., voire de la fin du siècle précédent. Un établissement agricole, constitué de plusieurs bâtiments sur poteaux, prend place à une vingtaine de mètres au sud de la batterie de silos, dont il est contemporain. Les opérations archéologiques menées dans un rayon d'un kilomètre ont par ailleurs mis en évidence trois autres habitats, chacun doté d'un nombre limité de structures de stockage probablement réservées à la consommation courante. La batterie de silos, pouvant accueillir jusqu'à 150 tonnes d'orge par saison, semble ainsi avoir rassemblé les surplus dégagés par l'exploitation des terres de ces différents établissements contemporains, traduisant une mise en commun du stockage des excédents à l'échelle d'un terroir. Si pour cet exemple un habitat (qualifié de "ferme-grenier" par S. Deffressigne) est directement associé à la batterie, les décapages réalisés autour du regroupement de 45 silos des Fontinettes à Cuiry-lès-Chaudardes (Aisne) n'ont livré aucun établissement rural à moins de 300 m (Gransar 2001, 342). De telles concentrations de silos sont aussi attestées en Maine-et-Loire, du VIe s. au IIIe s. (dès le VIe s. à Luigné [10 structures], puis en plus grand nombre durant les deux siècles suivant à Cizay-la-Madeleine et à Chacé [respectivement 20 et 47 silos] : Levillayer, Borvon, et Hunot 2013; Bossard et al. 2018) ; elles sont particulièrement bien documentées dans l'Aisne pour les IVe et IIIe s. (14 silos sont attestés aux "Sables Nord", à Limé, 45 aux "Fontinettes" [commune de Cuiry-lès-Chaudardes], 80 à Berry-au-Bac et 107 à Soupir "Champ Grand Jacques" : Gransar 2001, 2002), et plus ponctuellement dans le sud de la Seine-et-Marne et en Lorraine (59 silos au "Grand Canton Sud" à Marolles-sur-Seine et 24 au "Bois d'Échalas" à Ville-Saint-Jacques en Seine-et-Marne : Séguier 1995; Séguier et al. 2006-2007; 37 à Géric sur la commune de Terville en Moselle : Deffressigne et Landolt 2017). Comment expliquer le développement de ces concentrations de silos, à l'écart des centres de pouvoir aristocratique ? F. Gransar y voit la volonté de constituer des stocks de denrées, essentiellement des céréales, produits par plusieurs établissements et destinés aux échanges. Selon lui, une dégradation climatique, au IVe s., et donc une baisse des rendements serait à l'origine de restructurations socio-économiques et du développement de stocks spéculatifs au sein de silos (Gransar 2001, 79 et 374–75) . Néanmoins, les rythmes suivant lesquels se mettent en place ces stocks excédentaires étant différents en fonction des régions, la cause climatique n'est guère satisfaisante pour expliquer l'apparition des premières batteries de silos, certes moins fréquentes, dès la fin du VIe ou au Ve s.

Enfin, en Bretagne occidentale, pour cette première phase, les résidences de l'aristocratie fouillées à Paule (Côtes-d'Armor) et à Inguiniel (Morbihan) sont dotées d'un nombre particulièrement élevé de souterrains et de caves boisées – de quatre à une dizaine de structures enterrées. Bien qu'elles ne soient pas toutes contemporaines, ces aménagements révèlent également les volumes de stockage considérables que possédaient ces habitats particuliers, sans qu'il soit toutefois possible d'y percevoir la volonté de centraliser les produits récoltés par des établissements voisins. Contrairement aux batteries de greniers observées au sein d'autres résidences aristocratiques, les architectures souterraines, invisibles depuis l'extérieur, ne revêtaient en outre aucun caractère ostentatoire. À ce titre, l'habitat rural de Kermat III à Inzinzac-Lochrist (Morbihan) se démarque des autres établissements agricoles de la région : il concentre à lui seul la quasi-totalité des greniers connus pour La Tène ancienne en Bretagne. Près de 45 édifices de ce type ont été implantés le long de fossés ou de chemins aux VIe et Ve s., bordant à l'extérieur un enclos dépourvu de souterrain[5]. Le statut de ce site pose alors question : pourquoi rassemble-t-il autant de structures aériennes à l'extérieur d'un habitat enclos, dans un secteur où les productions agricoles semblent plutôt gardées sous terre, au cœur des établissements ? S'agit-il d'une exploitation fortement investie dans les productions agricoles, ou centralise-t-elle les récoltes des fermes voisines ? Il est difficile de trancher pour l'heure, cet exemple restant unique à l'échelle régionale.

3.2 *De La Tène moyenne à la fin de l'âge du Fer (milieu du IIIe s.–fin du Ier s. av. n. è.)*

La fin de l'âge du Fer (La Tène moyenne et finale) est marquée par de profondes mutations qui bouleversent l'organisation des sociétés gauloises. Dès la fin du IIIe s., des agglomérations ouvertes se développent en Gaule, suivies lors de la seconde moitié du IIe s. et au début du Ier s. par la fondation de véritables villes fortifiées, les *oppida*. L'alimentation des artisans résidant dans ces agglomérations, mais aussi des guerriers dont l'importance semble accrue au IIIe s. ou encore du personnel sacerdotal rattaché aux sanctuaires qui se multiplient en Gaule, suppose une productivité accrue de l'agriculture et l'ouverture de nouveaux marchés. Dans les campagnes, l'introduction de la meule rotative permet un rendement dix fois supérieur à celui de la meule va-et-vient qu'elle remplace dès la seconde moitié du IIIe s. en Gaule septentrionale. En parallèle, l'outillage métallique se diversifie, notamment avec l'apparition de la faux et du soc en fer (Marion 2013; Jaccottey et al. 2013). À la même période, les établissements ruraux investissent de nouveaux milieux, s'installant sur les plateaux et les cordons littoraux, en plus des vallées. Puis, au début de La Tène finale, soit au milieu du IIe s., les campagnes connaissent un développement sans précédent du nombre d'établissements ruraux, phénomène généralisé à l'échelle du territoire considéré (Blancquaert et al. 2012; Malrain, Blancquaert,

---

5 Information communiquée par J. Le Gall (Inrap), responsable de la fouille en 2015.

et Lorho 2013, 225). Quant aux productions agricoles, le spectre de plantes cultivées se spécialise dès le IIIe s. en s'orientant vers une intensification de la culture des céréales (blé amidonnier, orge vêtue et épeautre essentiellement), tandis que la pratique de la méture tombe en désuétude (Matterne 2000; Zech-Matterne et al. à paraître). Cette céréaliculture, alors propice au dégagement de surplus, nécessite des espaces adéquats pour entreposer et conserver les grains : comment évoluent les modalités du stockage alimentaire dans ce contexte ?

L'investissement de nouveaux terrains et la croissance du nombre d'exploitations agricoles fournissent une documentation abondante sur les structures de stockage, qui est aussi géographiquement mieux équilibrée que pour la phase précédente (fig. 4.8 à 10) – à l'exception des secteurs orientaux et méridionaux de l'aire considérée, peu renseignés.

D'une manière générale, les batteries de silos disparaissent au milieu du IIIe s. Dans le secteur occidental, ce même siècle constitue également une étape fondamentale : les derniers souterrains de la péninsule armoricaine semblent creusés durant cette période, ou au plus tard au début du IIe s. En tout état de cause, les derniers ouvrages de ce type sont abandonnés et comblés autour des années 150. En revanche, souterrains et caves boisées perdurent dans la Manche et le Calvados, où ils constituent des dispositifs essentiels au sein des établissements enclos de la Plaine de Caen. Composées d'ordinaire d'une salle creusée en fosse ou en sape et dotée d'un conduit d'aération, ces architectures enterrées sont généralement implantées à l'aplomb ou en périphérie de l'habitation, à laquelle elles sont reliées par un escalier d'accès. Elles apparaissent ainsi comme les structures dédiées à la conservation des aliments consommés par l'unité domestique. Aussi, les habitats de la Plaine de Caen sont le plus souvent pourvus d'une à plusieurs dizaines de fosses parallélépipédiques – destinées aux surplus dégagés lors des récoltes ? –, qui constituent l'autre composante fondamentale des établissements de cette région (Lepaumier et al. 2011; Bossard 2015; Bossard et al. 2018). En revanche, greniers et silos sont absents en Plaine de Caen. Si la conservation de légumineuses, attestées par la carpologie notamment au IIIe s., a parfois été supposée pour ces aménagements souterrains spécifiques à la Normandie occidentale, malgré le problème de germination qu'elle implique (Zech-Matterne, Auxiette, et Malrain 2013, 402), cette hypothèse doit être abandonnée : les établissements équipés de telles structures sont essentiellement tournés vers la céréaliculture (Lepaumier 2015). En tout état de cause, la Plaine de Caen, et probablement les secteurs voisins encore peu étudiés (Calvados, Manche) se caractérisent par le succès patent des aménagements de stockage souterrains à la fin de l'âge du

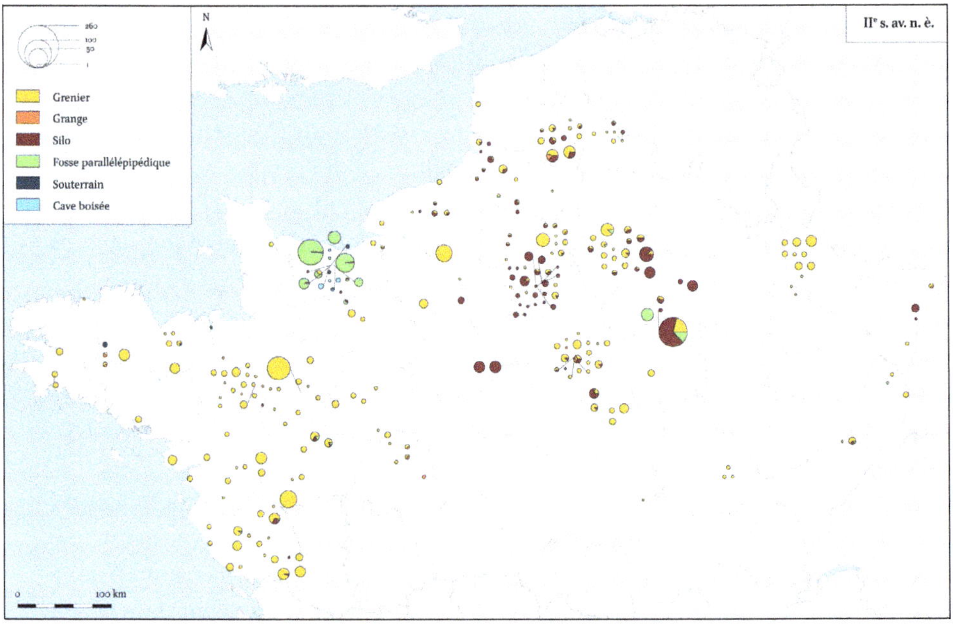

FIGURE 4.9 Répartition géographique des structures de stockage identifiées au IIe s. av. n. è.
S. BOSSARD

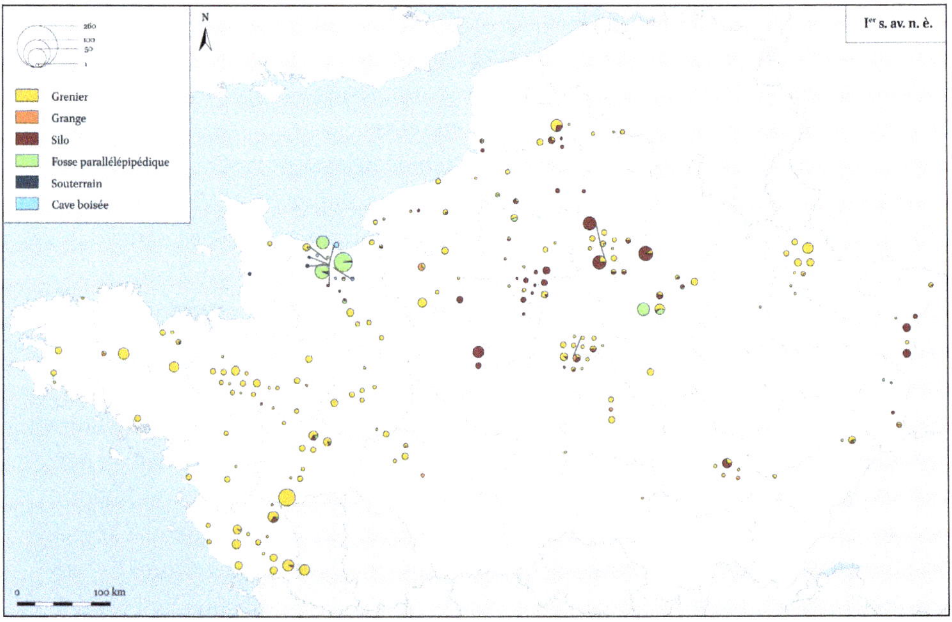

FIGURE 4.10 Répartition géographique des structures de stockage identifiées au Ier s. av. n. è.
S. BOSSARD

Fer. Celles-ci s'implantent au sein d'un réseau d'établissements ruraux prospères, et constituent un cas à part à l'échelle de la moitié nord de la France.

De fait, à l'exception de ce secteur particulier, les campagnes gauloises se couvrent de greniers aériens dès la fin du IIIe s., bien que le stockage en silos subsiste dans certains secteurs, soulignant l'existence de spécificités microrégionales. Les silos se retrouvent notamment dans le bassin parisien, en particulier en Plaine de France où ils sont globalement préférés aux greniers, ou encore dans le Nord de la France où ils coexistent avec les constructions surélevées. En dehors de la Plaine de Caen, les fosses parallélépipédiques ne sont attestées que ponctuellement et restent ainsi un mode de stockage – lorsqu'elles sont bien conçues à cette fin – anecdotique. Dans l'Ouest de la Gaule, en Bretagne et Pays de la Loire, le développement des greniers aériens, mode de stockage quasi exclusif à partir des années 250, est particulièrement remarquable, malgré la survivance de plusieurs silos et fosses de plan rectangulaire en Maine-et-Loire (Bossard et al. 2018).

En 2001, F. Gransar a constaté la disparition, à La Tène finale, des sites centralisateurs des récoltes et du stockage de masse dans le monde rural. Il suppose logiquement que les excédents produits dans les campagnes seraient directement acheminés, après la moisson, au sein des villes où ils seraient stockés avant d'être redistribués. Un passage de César (*Gal.* 7.32), qui mentionne "l'abondance de blé et d'autres vivres" puisée parmi les réserves de Bourges/ *Avaricum* pour nourrir l'armée romaine à l'issue du siège, appuie son hypothèse. F. Gransar précise qu'il "s'agirait alors de vaste batteries de structures, dont aucune n'est documentée à l'heure actuelle", vraisemblablement d'imposantes granges (Gransar 2001, 352–53, 2003, 209). Les découvertes archéologiques effectuées depuis ses recherches nourrissent ces réflexions à plusieurs titres.

D'une part, la diminution du nombre de structures de stockage par établissement rural à la fin de la période gauloise est certes réelle d'un point de vue général, mais ce constat doit être nuancé par la mise en évidence de certains sites particuliers. À partir du milieu du IIIe s., les établissements ruraux sont dans la majorité des cas pourvus d'un à plusieurs greniers, silos ou fosses quadrangulaires ; leur nombre excède rarement cinq à dix structures, avec des variations qui s'expliquent essentiellement par la taille de l'occupation et le nombre d'habitants à sustenter. Ainsi, l'enclos de taille modeste fouillé aux trois-quarts sur la ZAC du Ther à Allonne (Oise) a livré un silo et un grenier, à usage manifestement domestique (Beaujard, Malrain, et Auxiette 2006). Quelques établissements doivent néanmoins être considérés à part, car ils possèdent un nombre de silos, greniers ou fosses parallélépipédiques exceptionnellement important. Ainsi, dans le Calvados, les trois habitats du Parc sur l'Herbage à

Saint-Martin-des-Entrées et de l'Étoile I et III à Mondeville disposaient chacun d'une cinquantaine à une centaine de fosses parallélépipédiques ; à l'Étoile III, la capacité de stockage a été évaluée à plus de 300 m^3 (Besnard-Vauterin 2009, 45). Plus modestes, des batteries de 25 et 26 silos mises au jour sur les sites des Equiernolles à Bétheny (Marne) et du Fond du Petit Marais à Bucy-le-Long (Aisne) évoquent dans une moindre mesure les concentrations connues aux siècles précédents. En ce qui concerne les greniers, l'établissement de La Tène moyenne de Poulainville (Somme) illustre le fait que certaines exploitations continuent de concentrer des excédents probablement destinés à la redistribution auprès de sites voisins. Une douzaine de greniers, pour partie rassemblés auprès de deux grands bâtiments (destinés au fourrage ou aux animaux ?) ont pu accueillir 32 t de blé ou 38 t d'orge, soit des quantités dépassant les besoins des quatre unités domestiques, équipées par ailleurs de silos. F. Malrain note à ce propos que "la forte capacité de stockage est un atout non négligeable pour la vitalité économique du site. Elle offre une relative indépendance et permet de mieux maîtriser le marché, voire de jouer sur les cours lors des périodes de soudures" (Malrain et al. 2016, 154). Outre une fonction de centralisation et de relais, il est possible d'envisager que certains établissements ruraux se soient spécialisés dans des activités spécifiques nécessitant des stocks de céréales importants, telles que la meunerie. La mise au jour d'une dizaine de greniers, d'espaces de battage et de structures liées à la mouture au sein de l'enclos du Haut Kerrault à Laniscat (Côtes-d'Armor) plaide en ce sens (Roy 2009, 226–32).

Au IIe s., deux imposantes batteries de greniers, disposés suivant des alignements parallèles, ont été érigées au Clos des Primevères à Entrammes (Mayenne) et à Val-de-Reuil (Eure), sur le site du Chemin aux Errants (Guillier et al. 2015; Moreau, Granier, et Dubreucq 2015). Ces deux sites dédiés au stockage de masse de céréales (essentiellement du blé amidonnier à Entrammes, et de l'orge sur le second site) concentrent près de 80 et 40 greniers implantés à proximité de structures d'habitat, sous la forme d'un enclos observé dans son ensemble à Val-de-Reuil. Ces deux établissements, manifestement dédiés à la centralisation de récoltes de plusieurs établissements en vue d'une redistribution (à quelle échelle ?) constituent les exemples les plus récents de batteries rassemblant des dizaines de greniers en contexte rural, pour l'heure inconnus à la fin du IIe et au Ier s. av. n. è. Ils s'inscrivaient probablement dans des réseaux d'échanges à plus ou moins longue distance, avant l'apparition des premiers *oppida* ; les réserves d'Entrammes étaient d'ailleurs placées en bordure d'un axe routier important, et à proximité d'une agglomération fortifiée de la fin de l'âge du Fer, dont la chronologie et la relation avec la batterie de greniers ne sont pas clairement définies.

D'autre part, quels sont les témoins du stockage alimentaire au sein des agglomérations de la fin de la période gauloise ? À l'instar des habitats ruraux, les réserves domestiques étaient manifestement conservées au plus près des habitations, pour un approvisionnement quotidien, au sein d'un nombre réduit de structures de nature variable : silos et greniers sur le village ouvert d'Acy-Romance dans les Ardennes, ou encore fosses parallélépipédiques sur celui de Quimper (Finistère) (Lambot et Méniel 1993; Le Goff 2016a). En ce qui concerne les rares agglomérations fortifiées documentées par des opérations archéologiques, les observations sont similaires : l'emploi de greniers (Moulay, Mayenne) ou de silos (Condé-sur-Suippe/Variscourt, Aisne) en tant qu'annexes des maisons est attesté (Le Goff 2016b; Pion et al. 1997). L'emprise généralement réduite des opérations archéologiques menées sur ces agglomérations gauloises rend difficile la mise en évidence de bâtiments affectés au stockage en masse des denrées alimentaires provenant des campagnes environnantes. Toutefois, dans l'Ouest de la France, deux exemples mis au jour récemment illustrent manifestement un tel phénomène en contexte proto-urbain. La résidence élitaire de Paule, reconstruite dans le second quart du IIe s. av. n. è., se dote d'une vaste enceinte qui enserre l'agglomération nouvellement installée autour du noyau aristocratique. Les anciens souterrains de la résidence sont alors définitivement comblés, tandis que deux enclos, implantés au sein de la ville, renferment de puissants greniers et granges sur poteaux, édifices dont l'emprise au sol totale était supérieure à 300 m². L'accès à ces enclos vraisemblablement dédiés au stockage alimentaire était en outre protégé par un porche (Menez 2009, 387–91). L'*oppidum* de Vue, fondé sur la rive gauche de la Loire, devait participer activement aux échanges commerciaux. Au sein de cette agglomération fortifiée, a été mise au jour une batterie de 25 greniers organisés en deux rangées, associés à un habitat sur cour[6]. Dans ces deux cas, comme l'avait présumé F. Gransar, il s'agit bien de constructions aériennes, sous la forme de granges mais aussi de greniers, qui conservent les stocks alimentaires de l'agglomération. Il serait bien entendu imprudent de faire de ces deux exemples une généralité, mais ces découvertes récentes laissent espérer la mise en évidence au cours des prochaines années d'autres espaces consacrés au stockage au sein des villes gauloises, aspect encore méconnu de ces habitats groupés.

6 Fouille dirigée par Gérard Guillier (Inrap), 2014. Bossard et al. 2018.

Conclusion

Cette esquisse de l'évolution du stockage alimentaire dans les campagnes de l'âge du Fer, de la fin du Hallstatt à la fin de l'indépendance gauloise, met en évidence de multiples transformations dans les modes de stockage employés au sein des établissements ruraux au cours du temps, mais aussi des spécificités régionales, plus ou moins marquées en fonction de la période considérée. La fin du IIIe s. et le IIe s. constituent un moment décisif, où les pratiques évoluent de manière presque uniforme à l'échelle de la moitié septentrionale de la Gaule avec une généralisation du grenier aérien, à l'exception de quelques secteurs fidèles au stockage souterrain – en silo, en fosse parallélépipédique ou en salle creusée en sape ou en fosse. La place des granges reste difficile à appréhender, en raison de l'identification incertaine de ce type de construction. Cette enquête, appuyée sur les simples données numériques extraites de la base de données "datAFer2", mériterait d'être approfondie. Pour appréhender et quantifier d'une manière plus précise le phénomène du stockage, mais aussi mieux comprendre les établissements centralisateurs, il serait en effet indispensable de mesurer ou restituer les capacités de chaque structure de stockage et de chaque site, ainsi que d'évaluer les besoins réels des établissements, malgré les inconnues qui entravent ces raisonnements. Des études régionales, tenant compte tant des productions agropastorales attestées que des contraintes pédologiques ou climatiques, sont en outre nécessaires si l'on souhaite expliquer les choix préférentiels observés sur certains secteurs, privilégiant par exemple silos ou fosses parallélépipédiques.

À partir du IIIe et surtout du IIe s., le développement des agglomérations n'est pas sans conséquences sur le monde rural qui s'adapte aux nouveaux besoins introduits par l'apparition d'une tranche de population non productrice, mais aussi à l'ouverture de marchés à l'échelle des territoires contrôlés par les centres de pouvoir urbains. En revanche, il n'est pas prouvé que la Gaule ait été impliquée dès l'âge du Fer dans des échanges de denrées alimentaires à plus longues distances, notamment avec le monde romain, même si "les Gaulois septentrionaux ont eu les moyens humains et techniques de dégager des surplus céréaliers, et qu'une partie de ceux-ci a peut-être été destinée à alimenter des villes méditerranéennes" (Gransar 2001, 360).

CHAPTER 5

De nouvelles formes de stockage de céréales à l'époque romaine en Gaule : quels changements, avec quel(s) moteur(s) ?

Alain Ferdière, avec la collaboration de Véronique Zech-Matterne** et Pierre Ouzoulias****

La présente contribution fait pour moi suite à un article concernant les greniers ruraux en Gaule récemment publié (Ferdière 2015). L'objectif présent est de tenter d'approfondir la question au-delà de ce que j'avais pu faire alors ;

- par une sélection plus sévère des types de greniers retenus, en les limitant aux deux les plus sûrs et les mieux représentés, leur conférant ainsi une plus grande valeur statistique ;
- par une révision du mode de calcul des capacités en fonction des surfaces, ici plus strict ;
- par l'ajout d'un nombre d'exemplaires relativement important, pour les deux types conservés ici ;
- par l'examen plus approfondi, enfin, de ces nouvelles évaluations de capacités de stockage et, partant, par l'évocation – délicate – de l'évaluation des rendements.

Il s'agit de réviser ici la question des capacités de stockage céréalier apparemment assez massives, essentiellement en milieu rural, en Gaule romaine, surtout dans sa partie septentrionale (pour l'essentiel au Haut-Empire, parfois au Bas-Empire).

Pour la période antérieure gauloise (La Tène Finale), on observe que des modifications importantes sont déjà intervenues avant la Conquête, sans doute dès le IIe s. av. n. è. et la période de La Tène C, avec l'abandon de la technique de stockage-conservation des céréales en espace souterrain confiné – les silos –, alors systématiquement remplacée par les greniers aériens à plancher

* Laboratoire Archéologie et Territoires, UMR CITERES, Tours ; ferdiere@club-internet.fr.
** CNRS, UMR 7209 AASPE, CNRS/MNHN/Sorbonne universités ; zech@mnhn.fr.
*** CNRS, UMR 7041 ArScAn, CNRS/Université Paris I Panthéon-Sorbonne/Université Paris Nanterre/MCC ; pierre.ouzoulias@mae.u-paris10.fr

DOI:10.1163/9789004389045_007

surélevé sur poteaux plantés, généralement de petite capacité (voir Chapitre 4 de ce volume, avec bibliographie antérieure). On verra plus loin les possibles comparaisons entre les capacités de stockage rural de céréales à la période laténienne et celles de la période romaine examinées ici. On observe ensuite, entre la Conquête et le milieu/seconde moitié du Ier s. de n. è., le passage de greniers aériens en bois, en général de petite taille (type 1 de Ferdière 2015) – existant à la période laténienne – à de vastes et puissants greniers maçonnés (voir Ferdière 2015).

1 Les grands greniers ruraux gallo-romains

Je ne reprendrai donc ici en ligne de compte, pour l'essentiel, que les types de greniers qui me semblent pouvoir être indubitablement identifiés comme tels parmi les bâtiments des parties d'exploitation des *villae*, soit seulement trois des sept types que j'ai récemment proposés pour les greniers ruraux en Gaule romaine (Ferdière 2015) : le type à contrefort et celui à piliers internes (nefs), ainsi que, le cas échéant, le type à murets parallèles (plancher surélevé). J'ai ajouté à mon premier inventaire un certain nombre d'exemplaires – non moins de 66[1]– pour ces deux types.

Précisons que, pour le Nord de la Gaule et notamment le Bas-Empire, Paul Van Ossel (1992, 154-55) proposait déjà de compléter la typologie de Pat Morris (1979) avec trois séries représentées dans ma typologie initiale, le type à contreforts, le type à plan basilical à piliers et le type tour.

Il est en outre à noter que, si les exemplaires retenus concernent en grande majorité des sites ruraux, des exploitations agropastorales en général de type *villa*, j'ai pensé utile de mentionner – ne serait-ce qu'à titre de comparaison – quelques cas présents dans des agglomérations secondaires, voire un chef-lieu de cité tel qu'Avenches.

– Le type à contreforts (*Strebepfeilerbau* ; Fig. 5.1a)[2] est bien représenté dans tout le Nord de la Gaule et par exemple assez bien présent en Picardie (Bayard et Lemaire 2014, fig. 30-31) où il s'agit pratiquement du seul type de notre typologie clairement identifié dans cette région. S'il s'agit de granges à contreforts, prises en compte ici, c'est le sous-type b (Ferdière, Gandini, et

1 Numérotés, à la suite des 156 items de l'article de 2015, de 157 à 222. Ces n°, qui figurent ici dans les cartes et autres documents, renvoient au corpus de greniers commun aux différentes contributions de la présente publication, sous forme d'un tableur Excel (.xlsx), accessible en ligne : <https://hal.archives-ouvertes.fr/hal-01812097>.

2 Type 2 *in* Ferdière 2015, 12–19.

a

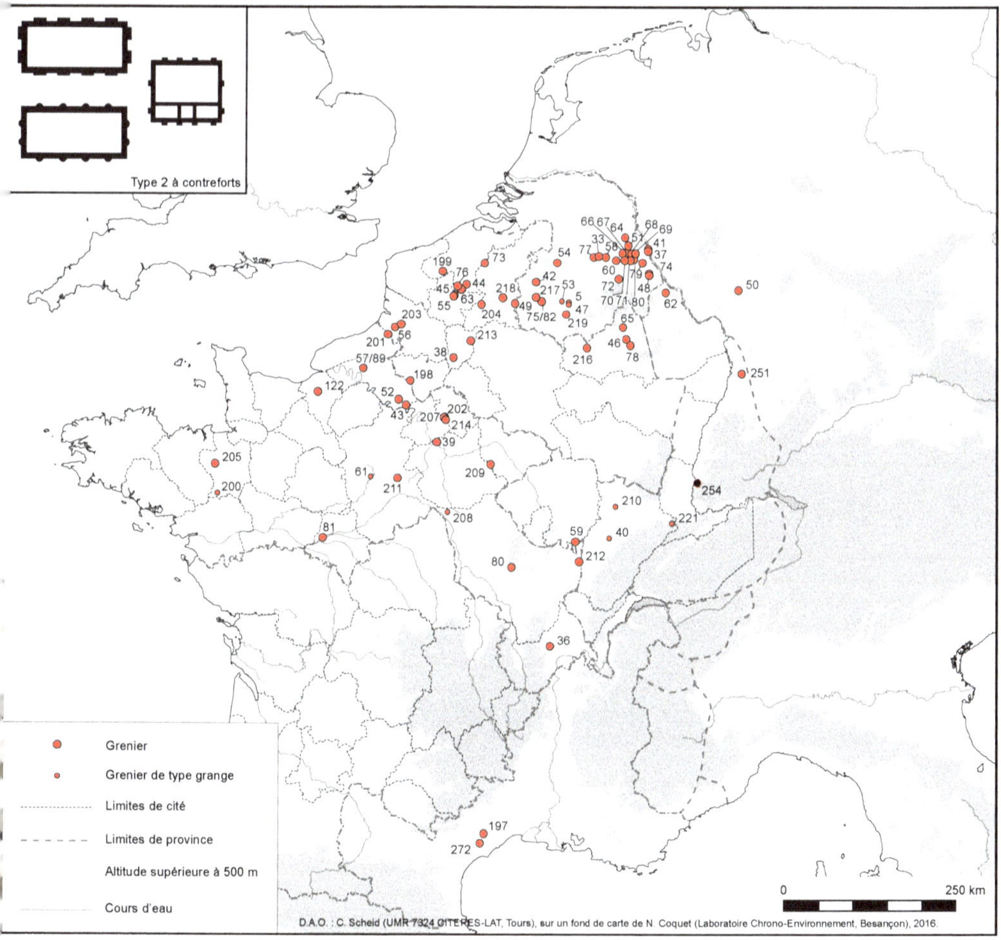

FIGURE 5.1 Cartes des deux types de greniers pris systématiquement en compte ici, mises à jour (incluant les types mixtes) : a) pour les greniers de type 2, à contreforts ; b) (page suivante) pour les greniers de type 3, à piliers internes
DAO CORINNE SCHEID, LAT/UMR CITERES, TOURS

b

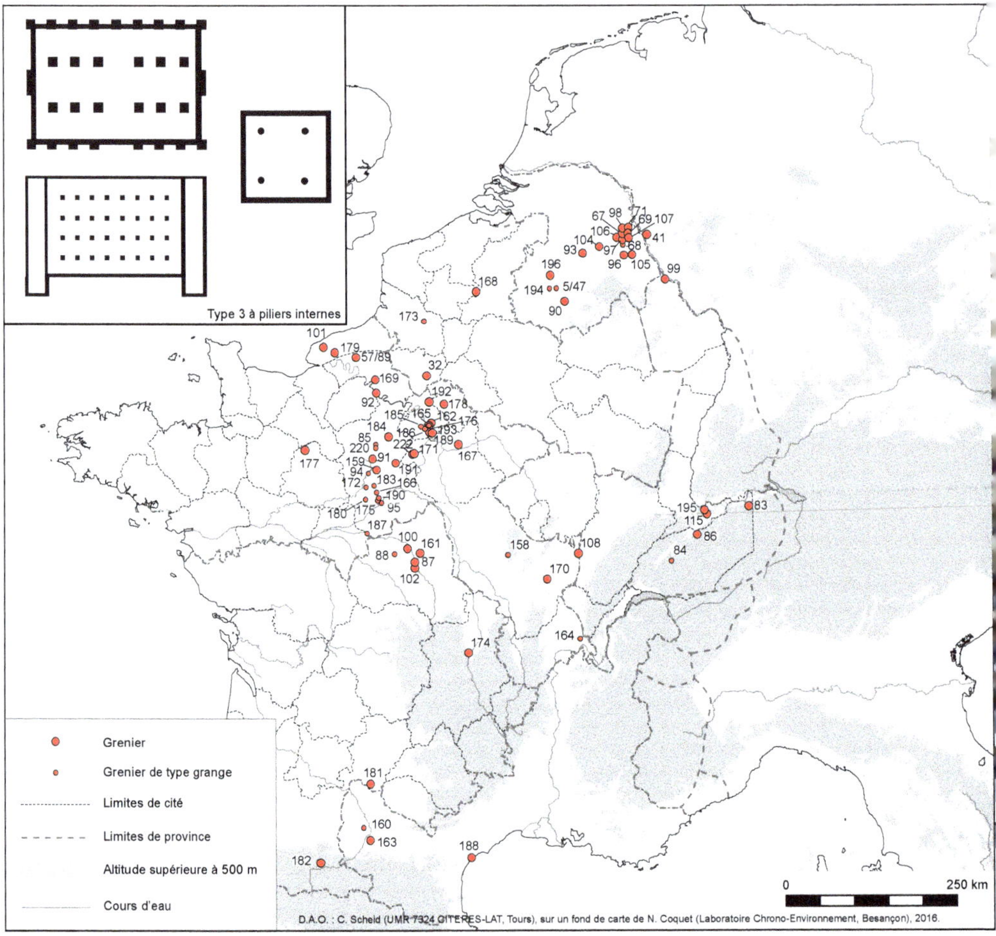

FIGURE 5.1 (cont.)

Nouvel 2017, fig. 13) qui est concerné[3]. Pour la présente étude, sont également inclus dans cette catégorie les greniers présentant à la fois des contreforts et des piliers internes.

- Le type à piliers internes et plusieurs nefs (dont plan basilical, *aisled buildings* ou *aisled houses*… ; Fig. 5.1b)[4] a – notons-le – longtemps été interprété comme partie résidentielle de *villa*, alors qu'il s'agit à mon avis non d'habitation mais de grands greniers : cette conviction s'inspirait de la vieille typologie des *villae* du *Manuel* d'Albert Grenier (1934, 802-5,"le type basilical", en fait repris de Haverfield 1924, 227 sqq. : c'est le "*aisled house/building*" de la bibliographie anglaise ; Ferdière 1988a, 168-169, fig. p. 169, 3 ; Agache (1978, 282) évoquait déjà la possibilité qu'il s'agisse de "granges" au moins pour certains exemplaires, suivant en cela Applebaum (1972, 133), mais en Picardie pour des plans à trois nefs séparées par des murs et non des rangées de piliers, plans que je n'ai pas retenus ici). Cette typologie a été de nombreuses fois reprise depuis, notamment avec les travaux, discutables, de John T. Smith (1963, 1997, 37-40, fig. 7 ; voir A. Smith et al. 2016, 58-60, fig. 3.12), surtout pour la *Britannia*, puis par exemple par Hermann Hinz (1970, fig. 6). Dans les exemples le plus souvent mentionnés par ces auteurs, je propose de voir de grands greniers à piliers internes définissant trois nefs : *villae* de Spoonley Wood, Knowl Hill, Ickleton, Chesterford, Petersfield Stroud ou West Blanchinton (GB), ou encore Kölstein (DE)[5] et bien sûr Seeb (CH) et Maulévrier (FR)[6] … Assez tôt, cette interprétation a été avancée (cf. Van Ossel 1992, 154), sur le modèle de grands *horrea* urbains connus en Gaule du Nord (Belgique et Germanies), tels que ceux de Trèves ou Cologne[7]. Il est à noter que ce type est apparemment totalement absent de toute la région, pourtant bien documentée, de Picardie (Agache 1978 ; Bayard et Lemaire 2014, fig. 30-31), où le type à contrefort (*supra*) est en revanche plutôt bien représenté. Et rappelons que ce même plan à piliers internes est généralement interprété, en tout cas en Narbonnaise, comme chai viticole. Mon type 4 (Ferdière 2015, 26-29), précédemment défini comme "à plan basilical", est assez flou (cf. *infra*) et recoupe en partie celui ici considéré, dont les piliers sont disposés en lignes parallèles, définissant assez couramment trois nefs, pour une élévation basilicale où seule la nef centrale disposerait sans doute d'un étage.

3 Symboles de plus petite taille sur les deux cartes mises à jour.

4 Type 3 *in* Ferdière 2015, 19–26.

5 Exemplaires (dont britanniques) non pris en compte ici.

6 Ces deux derniers sites intégrés ici en tant que greniers : cat. 83/103 et 179.

7 À noter que de tels bâtiments à piliers internes, possibles greniers, sont assez couramment présents parmi les bâtiments d'exploitation de *villae* dans d'autre provinces, par exemple danubiennes ou encore en Grande-Bretagne (cf. par ex. Smith 1997, 245, fig. 67, etc.).

Quand il s'agit de granges – bâtiment en principe plurifonctionnel – à piliers internes, concernées ici, ce sont les sous-types a et e2 (Ferdière, Gandini, et Nouvel 2017, fig. 13) qui sont concernés[8].

Il est nécessaire de faire observer que, du moins à partir de la photographie aérienne, méthode d'acquisition de données très largement mise en œuvre, les contreforts accolés aux murs peuvent, globalement, être moins facilement reconnus que les piliers internes.

Sur les sept types retenus pour mon précédent article (Ferdière 2015)[9], les cinq autres n'ont pas – sauf exception – été repris ici, car moins représentatifs ou connus en nombre insuffisant pour permettre des considérations générales de capacité. Cependant, ces autres types seront mentionnés :

- soit s'ils concernent des établissements qui comportent par ailleurs des greniers des deux types pris en compte, afin d'essayer d'approcher les capacités globales de stockage des établissements, et non celles de chacun des greniers, individuellement ;
- soit s'il s'agit de types "mixtes", comportant à la fois des contreforts et des piliers internes (voir *infra*).

Notons cependant que l'un des types non examinés ici – mon type 7, le grenier-tour ou tour-silo (*Speicherturm*) – a fait l'objet d'une attention particulière dans le travail de P. Van Ossel (1992, 155), qui en recensait dès lors non moins de 10 exemplaires[10], surtout au Bas-Empire, pour le seul Nord de la Gaule, notant qu'ils étaient souvent associés sur les mêmes sites à d'autres types de greniers, comme c'est le cas à Voerendaal, Cologne "Müngersdorf", Niederzier V, Echternach ou Wasserbillig I[11] : à partir d'un premier exemple proposé sur la *villa* de Mayen, ce type a en effet fait l'objet d'une abondante bibliographie, suggérant notamment leur lien probable avec l'approvisionnement en grain du *Limes* rhénan au Bas-Empire, une majorité des greniers de ce type, incluant quelques exemplaires fortifiés, datant de cette période. Le grenier-tour, existant déjà au

8 Symboles de plus petite taille sur les deux cartes mises à jour (Fig. 5.1a et b).

9 Il est à noter que les greniers sur lesquels a raisonné Lars Blöck (2011-2012, cf. fig. 2 et tabl. p. 98–99 ; cf. sa contribution dans le présent volume) – et notamment donc Heitersheim – sont à mon sens plutôt atypiques et à ce titre non retenus ici, à l'exception de Biberist et Voerendaal.

10 J'en considérais sept dans mon étude (Ferdière 2015, 36–38), auxquels je pourrais donc aujourd'hui ajouter neuf exemplaires, dont trois signalés par P. Van Ossel (1992) non encore pris en compte.

11 Ici cat. 33, 41, 68, 46 et 78.

Haut-Empire, aurait ainsi été à l'origine de la plupart des *burgi* tardifs (Bechert 1978).

Les raisons du rejet, dans la présente étude, des 5 autres types de grenier précédemment définis (2015) sont donc, outre leur plus faible représentation qui oblitère leur représentativité statistique, de divers ordres :

- pour le type 1 (à plateforme sur rangées de poteaux), il ne s'agit que d'une forme "de transition" reprenant les modèles de greniers à plateforme surélevée sur poteaux de la période laténienne ;
- pour le type 4 (à plan basilical, cf. *supra*), il s'agit d'un plan architectural trop polyvalent pour permettre une identification à peu près assurée comme grenier[12] ;
- pour le type 5 (à murets parallèles internes), il a été examiné récemment à l'occasion de deux contributions au colloque AGER de Clermont-Ferrand (Fouillet et Morillon 2017 ; Poux 2017 ; cf. aussi Poux et Borlenghi 2016), et il ne m'a donc pas paru utile d'y revenir, d'autant que les exemples ruraux restent relativement rares[13] ;
- pour le type 6 (à caissons maçonnés), c'est seulement leur faible nombre, ainsi que l'ambiguïté de certains exemplaires, qui ont occasionné ce rejet ;
- pour le type 7 enfin (grenier-tour), le type apparaît trop hétérogène (en tout cas dans les exemples sélectionnés par Pellegrino, Mauné, et Mathieu 2017) et mal caractérisé, et le plan est ici aussi assez ambigu : l'identification comme grenier reste donc souvent discutable (sauf présence de piliers et/ou de contreforts, exemples alors pris en compte).

On remarque que les greniers maçonnés des divers types définis précédemment ne sont pas particulièrement représentés dans les très grandes *villae* à pavillons multiples étudiées il y a quelques années (Ferdière et al. 2010), notamment pour les régions où ce type de *villae* est bien représenté : une petite dizaine de ces *villae* en comportent, sur les quelques 130 enregistrées alors ; et un seul grenier y est en général répertorié, surtout du type à piliers internes (dont des granges). On peut donc se demander si, sur ces très vastes établissements, le nombre de petites structures de stockage ne compenserait pas en

12 Comme je le notais dans mon précédent article (2015), c'est pourtant le seul plan de bâtiment qui puisse être proposé dans plusieurs cas pour les multiples *villae* connues en Picardie : il est caractéristique à ce titre que le récent bilan de Didier Bayard et Patrick Lemaire (2014, 49–72) ne réserve pas de développement spécifique au "grenier" parmi les divers types de bâtiments d'exploitation rurale reconnus dans cette région.

13 On mentionnera toutefois l'exemplaire de Voerendaal, accompagné d'autres types quant à eux retenus ici.

quelque sorte l'absence de structures de grande superficie. Néanmoins la très grande *villa* à pavillons multiple de la "Mare aux Canards", récemment fouillée à Noyon (De Muylder 2014) a fait l'objet d'un échantillonnage carpologique extensif. Près de 300 prélèvements ont été effectués ; 9256 restes en ont été extraits, dont 1858 restes carbonisés, qui révèlent l'absence de concentrations de grains pouvant évoquer des résidus de stocks céréaliers, ainsi que la très faible proportion de sous-produits résultant du traitement de récoltes (voir l'étude de Véronique Zech-Matterne dans Malrain et al. 2017). Alors que cette *villa* se trouve dans une zone de production céréalière vouée aux blés nus, l'amidonnier et l'orge y constituent les céréales dominantes, et en vertu des faibles effectifs recueillis (moins d'un reste carbonisé au litre, pour une quantité de sédiment prélevée équivalant à un volume global de 2040 litres), la production céréalière y apparaît des plus discrètes : effet de la taphonomie, ou réalité économique ? Étant donné le nombre et la variété des contextes échantillonnés, les résultats peuvent être considérés comme significatifs, et certainement plus étoffés que bon nombre de séries étudiées à l'échelle régionale. Enfin, il paraît utile d'attirer l'attention (cf. Tableau général) sur les sites de Niederzier (forêt de Hambach, dans la cité de Cologne), avec non moins de 13 établissements différents qui présentent d'1 à 3 greniers, à contreforts et/ou piliers internes[14].

• • •

Les deux cartes ainsi mises à jour (Fig. 5.1a et 5.1b) pour les types 2 et 3 (*supra*) sont assez suggestives et méritent quelques commentaires.

- On note tout d'abord, clairement, qu'une large moitié sud de la Gaule est très peu concernée pour les deux types, et qu'il en est globalement de même pour le Nord-Ouest (Armorique) ainsi que pour l'Est (peut être à l'exception de la Suisse pour le type 3). On pourrait certes invoquer des carences de l'enquête et donc des lacunes du corpus pour ces régions, mais le contraste est suffisamment flagrant pour laisser penser qu'il s'agit de la réalité[15].
- On observe ensuite quelques rares secteurs où les deux types sont apparemment bien représentés et donc concurremment : surtout le centre (sud-est) de la province de Germanie Inférieure, avec notamment la cité des Ubiens

14 Dont non moins de 9 greniers sont standardisés à 200 m² de surface interne (cf. *infra*).

15 Ces cartes ne prennent en effet pas en compte 5 ou 6 sites pour le type 3 et un pour le type 2 dans le Centre-Est (Nouvel 2004), ce qui reste assez insignifiant sur les nombres de sites concernés pour ces deux types.

(Cologne), et, dans une moindre mesure, l'ouest de la province de Belgique (au nord de la Seine).

- Enfin, dans certains secteurs, l'un des deux types apparaît clairement dominant : c'est le cas du type à contreforts (type 2) pour la Picardie et en général le nord de la France[16], mais, à l'inverse, du type à piliers internes (type 3) pour les cités d'Île-de-France et du Centre-Val-de-Loire (dont les Carnutes, avec la Beauce, et les Bituriges Cubes)[17].

Il reste en outre nécessaire de bien préciser que ces deux cartes ne sont représentatives – et encore dans une certaine mesure – que de la présence des deux types ici considérés de bâtiments agricoles, interprétés comme greniers. Ces documents ne pourraient en aucun cas être assimilés à une cartographie des régions et cités des Gaules les plus fertiles en matière de production céréalière. Dans de nombreux secteurs, d'autres types de bâtiments – y compris non encore identifiés comme greniers – sont certainement mis en œuvre : on le constate de manière patente, par exemple, pour la Picardie, avec sans doute les bâtiments de plan basilical (Agache 1978 ; Bayard et Lemaire 2014), ou encore avec la Beauce, où s'impose clairement la grange de plan stéréotypé (Ferdière, Gandini, et Nouvel 2017).

2 Les capacités de stockage

Il est ici nécessaire de changer de mode de calcul de capacité des greniers, par rapport à celui proposé dans mon précédent article (2015, 39), en vérité trop simpliste. Tout en conservant les évaluations proposées par Véronique Matterne, Jean-Hervé Yvinec et Dominique Gemehl (1998)[18] pour le petit grenier urbain d'Amiens, soit une épaisseur de 30 cm de grain, il faut admettre qu'il est nécessaire de ménager des tas séparés dont l'étendue soit compatible avec les nécessités de remuer le grain pour l'aérer régulièrement (Diffloth 1917) : soit des tas – disons – de 5 × 5 m au maximum à la base, qui, s'ils ne sont pas dans des espaces coffrés, nécessitent en outre des pentes latérales à environ 45° (mais voir Chapitre 3 dans ce volume pour des estimations plus basses). Avec

16 Quelques exemplaires de l'ex-région de Picardie resteraient à intégrer dans ces cartes, selon Bayard et Lemaire 2014, mais ne feraient qu'accentuer ce contraste.

17 Région, notons-le, où une proportion représentative de ces greniers à piliers internes sont en fait de petite dimension, sous la forme de la grange de plan stéréotypé (Ferdière, Gandini et Nouvel 2017).

18 Évaluations possibles grâce à l'excellente conservation du grain brûlé, sur toute l'épaisseur du dépôt, dans ce grenier incendié ; cf. aussi Matterne 2001, 150.

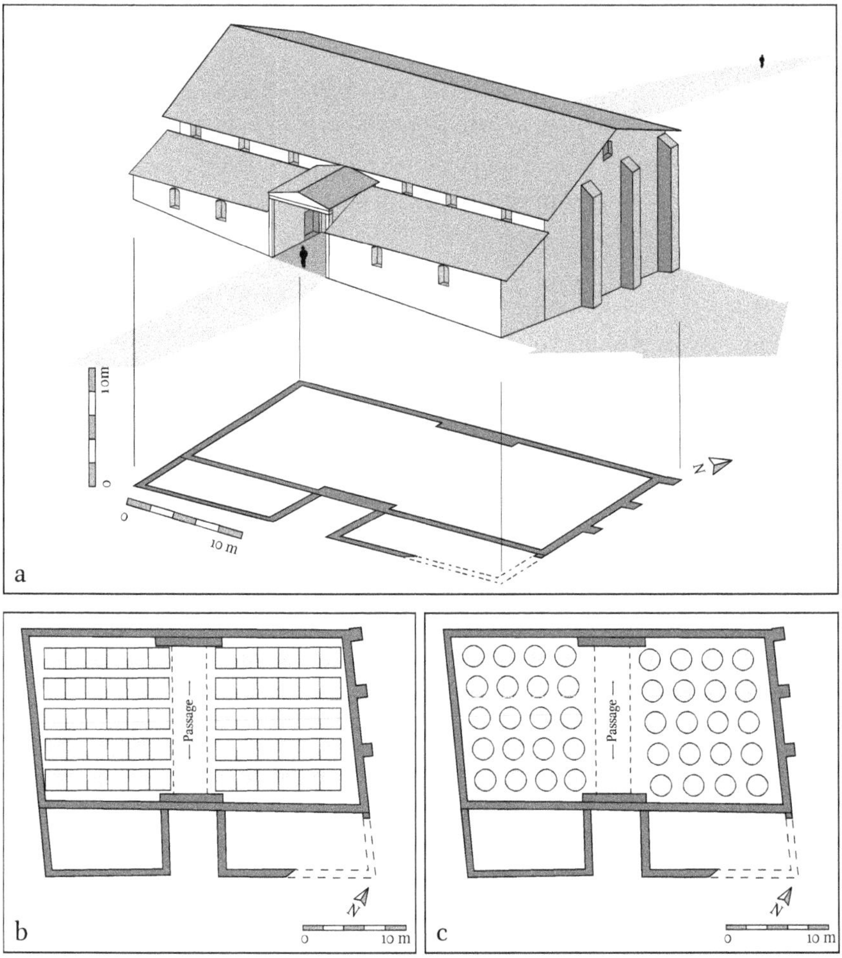

FIGURE 5.2 Le grand grenier à plancher surélevé sur piliers ou lambourdes à Alle : a) proposition de restitution axonométrique de l'élévation ; en bas, deux propositions de mode de stockage du grain, b) en compartiments allongés ou c) en tas de 2 m de diamètre
D'APRÈS DEMAREZ ET AL. 2010, FIG. 359, 362, 363

les passages nécessaires entre les tas ainsi qu'à la périphérie, le long des murs du grenier[19], on enregistre donc une perte de capacité de stockage par niveau de l'ordre de 1/5 à 1/4 de volume de grain par rapport à une couche continue sur toute la surface du grenier…

---

19 À Amiens, selon les dimensions du grenier, sans doute un seul tas (sauf petits lots d'autres graines localisés), avec passage périphérique (observé à la fouille en bordure d'amas de grain brûlé).

Notons que c'est une épaisseur de 0,30 m, pour des tas sans caissons, avec des allées de circulation d'1 m de large, qui est retenue pour la restitution du grand grenier d'Alle, dans le Jura Suisse (ici cat. 125/221 ; Demarez et al. 2010, 393-94, fig. 362 ; cf. fig. 5.2 du présent chapitre)[20]. Pour les grands greniers de Panossas (ici cat. 314), Matthieu Poux propose quant à lui 0,40 m d'épaisseur, en mentionnant une proposition d'1 à 1,40 m pour un grenier urbain de Vienne (ici cat. 274)[21]. Cette épaisseur de 0,30/0,35 m envisagée ici reste donc très inférieure à celle proposée par Lars Blöck (2011-2012, 86), soit 0,90 m, qui me paraît en effet incompatible avec cette nécessité de remuer le grain[22].

En revanche – tout en tenant donc compte de cette perte de capacité due aux nécessités d'aération du grain (à la pelle) –, je maintiendrai, pour les deux types de greniers réexaminés ici (à contreforts et à piliers internes), la proposition de deux niveaux de stockage superposés, sur toute la surface interne au sol[23].

J'ai donc créé un "bâtiment-modèle" (Fig. 5.3) – certes un peu théorique, par définition – de grenier massif, à contreforts et piliers internes, ici en une seule file centrale (type mixte 2-3), pour un stockage sur planchers surélevés, sur deux niveaux, de surface interne de 25 × 13 m (soit 325 m²). Ce bâtiment-modèle théorique, pour le fondement des calculs, est rectangulaire (26 × 14 m), avec des murs épais (0,50 m) et contreforts aux angles et sur les 4 côtés (3 sur les longueurs, 1 sur les largeurs) ; il présente un sous-sol bas (H = 0,80 m), avec un ligne de 3 forts piliers centraux supportant, aves les murs, de solides lambourdes pour un plancher surélevé ; sur un des grands côtés, une rampe-escalier permet d'accéder, par une large porte, à ce niveau de stockage ; les piliers axiaux supportent en élévation des piliers maçonnés ou plutôt de forts poteaux de bois qui reçoivent eux-mêmes les charpentes pour un étage (second

20 Et tas de 0,80 à 1 m de haut après pelletage, avec la nécessité d'aménager des passages ; c'est cette épaisseur de 0,40 m qui a également été retenue pour le grenier de la *villa* de Champion (*infra* : Van Ossel et Defgnée 2001, 232).

21 Pour Panossas, Matthieu Poux s'appuie aussi sur V. Matterne (2001, 150), qui s'appuie elle-même sur les agronomes modernes et contemporains (antérieurs à l'industrialisation et la mécanisation de l'agriculture) et l'ethnographie; cf. Diffloth 1917 ; Sagnier 1920 ; Chancrin et Dumont 1921, 782-83 ; Cuvillier 1931. Vienne : Helly 2014, 143.

22 La proposition de Lars Blöck est fondée sur des données ethnographiques ou récentes allemandes (Krünitz 1788) ; comme pour Amiens (n. 19, *supra*), il faudrait discuter de la pertinence de ces "modèles" modernes ou ethnographiques par rapport à cette question de capacité des greniers, mais ceci méritera de plus amples débats par la suite… Et voir l'interrogation de Stefan Wenzel sur Researchgate du 11 août 2016 : <https://www.researchgate.net/post/Does_anyone_knows_up_to_which_height_cereals_can_be_stored_on_raised_floors> (consulté le 7 juin 2018)

23 Le premier niveau lui-même déjà sur plancher surélevé, au moins sur lambourdes.

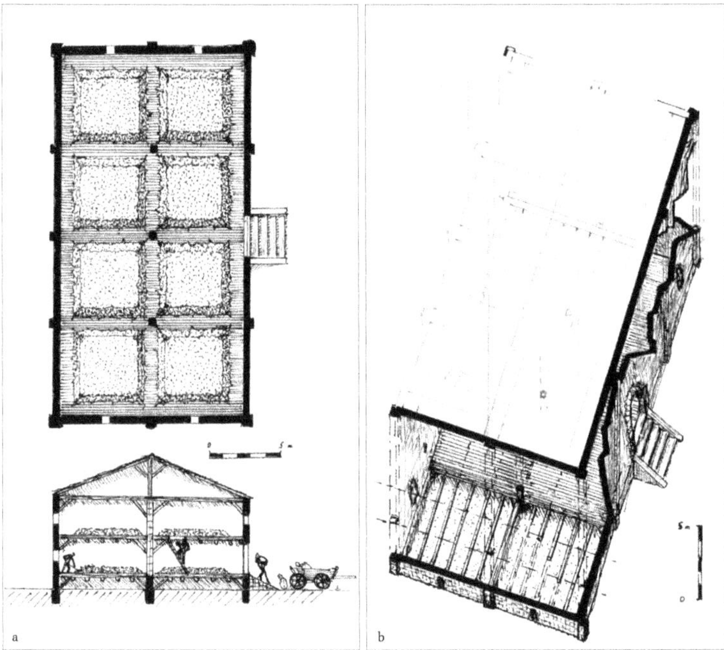

long. ext.	26 m
larg. ext.	14 m
haut. tot. sous charpente toit	6 m
long. int.	25 m
larg. int.	13 m
surf. int.	**325 m²**
n. de niveaux	2
vol. max. stockable théorique	113,75 m³
n. tas par niv.	8
dim. tas à la base	5x5 m
haut. tas (pyram.)	0,35 m
capacité max. en tas	60,64 m³

c

FIGURE 5.3 Modèle-type de grenier gallo-romain (type "mixte" ici, à contreforts et piliers internes), proposé pour servir de base de calcul de capacité : a) plan et coupe ; b) vue axonométrique, avec les parties supérieures (2e niveau de stockage et couverture) ici seulement suggérées ; c) tableau des données métriques du modèle
A. FERDIÈRE

niveau à plancher, de stockage) puis la toiture (à fermes) ; l'accès à ce second niveau de stockage, dont la superficie est égale à celle du premier, ne peut se faire que par une échelle, avec trappe ; la hauteur des deux étages a été établie à seulement 2,5 m, suffisante sous plafond pour travailler, y compris sur les tas de grain de 0,35 m de hauteur. Des ouvertures sont prévues aux deux étages, ainsi qu'au sous-sol, pour assurer l'aération (Fig. 5.3a, b et c).

J'envisage à chaque niveau 8 tas de 5 × 5 m à la base, séparés et entourés par des allées de circulation larges d'1 m, et constitués de grain sur 0,35 m d'épaisseur, avec des pentes de 45° sur les quatre côtés, soit des troncs de pyramides (et donc sans caissons de maintien) d'un volume de 7,58 m³ chacun.

Le volume total alors stocké par niveau, en 8 tas, est précisément de 60,64 m³, contre les 113,75 m³ obtenus si la totalité de la surface du grenier avait été utilisée (selon la base de calcul précédemment adopté : Ferdière 2015, 39) : la "perte" de capacité est donc dans ce cas de 53,11 m³, soit précisément 46,69 %, presque la moitié. Pour le nouveau calcul des capacités de stockage des exemples de greniers retenus, je pense par conséquent pouvoir m'autoriser à simplifier ceci[24], en considérant une perte moyenne de 45 % par rapport au volume calculé sur toute la surface utile. Pour les quelques greniers intégrés dans la présente étude et celle de Lars Blöck (2011-2012, 92–99, tabl. p. 98–99 ; voir aussi son Chapitre 1 dans le présent volume), les résultats seront donc sensiblement différents.

Le calcul de capacité (en mètres cubes) a donc ensuite été fait ici sur cette base[25], en prenant en compte deux niveaux de stockage, tant pour les greniers de type 2 (à contreforts) que de type 3 (à piliers internes[26]). Ce résultat est alors noté par site concerné, en additionnant les capacités des différents greniers si le site en présente plusieurs, et en considérant arbitrairement qu'ils ont fonctionné de manière contemporaine[27] (sauf information explicite contraire dans la publication).

Comme indiqué *supra*, j'ai par ailleurs cru bon de conserver dans le tableau ainsi mis à jour un certain nombre de sites urbains (chefs-lieux ou agglomérations secondaires), à titre de comparaison.

Sur cet échantillon de 148 sites ici pris en compte (pour 188 greniers), les capacités totales de stockage de grain – en admettant que tous les greniers d'un même site possédant plusieurs greniers ont fonctionné ensemble[28] – s'étagent, selon ce mode de calcul (Tabl. 5.1), sur une gamme très vaste : de 6 m^3 pour le plus modeste (cat. 253, Hardivilliers) à non moins de 1351,5 m^3 pour le plus important (cat. 108, Saint-Aubin, cf. Fig. 5.8).

Le graphique montrant les différentes capacités par site (Fig. 5.4a et 5.4b) révèle une progression continue assez régulière de ces capacités[29] pour la grande majorité des sites pris en compte, au-delà de laquelle en ne perçoit nettement que cinq ruptures :

---

24 Compte tenu notamment d'une certaine irrégularité des tas ainsi que du tracé des allées de circulation, pour un stockage sans coffrages.

25 Je dois donc, en toute honnêteté, signaler ici que j'ai pour la publication entièrement revu les bases de calcul de ces capacités et donc les résultats par site maintenant pris en compte ; les propositions initiales, au moment la communication à l'atelier de décembre 2016, étaient en effet entachées d'une importante erreur de calcul sur le volume ainsi stocké par niveau de grenier, qui se répercutait sur l'ensemble des estimations proposées par sites ; j'ai en outre corrigé, notamment grâce à Stéphane Martin, les superficie de plusieurs greniers, ce qui modifie aussi les résultats finaux…

26 Contrairement à mon article de 2015, où ils étaient considérés comme à un seul niveau.

27 Valeurs totales arrondies au demi-m^3 le plus proche.

28 Pour certains sites, on a pu distinguer deux phases, séparées ici.

29 Progression de 6 à 134,5 m^2 (rapport de 1 sur 22,4), avec 121 sites sur 148 (81,75 % des sites concernés).

TABLEAU 5.1 Tableau ordonné des capacités estimées, selon le mode de calcul proposé, pour l'ensemble des établissements ruraux à greniers pris ici en compte

N° site	Capacité en m3	N° site	Capacité en m3	N° site	Capacité en m3	N° site	Capacité en m3
253	6	186	43	188	75,5	264	130
165	7	78	43	55	76	134	130
216	11,5	130	44	222	77	05,	131
133	13,5	76	44,5	171	77	47	
69	13,5	51	47	172	77	50	134,5
81	17	132	47,5	72	77	169	128,5
127	19	52	48,5	184	77	176	128,5
57,	22	189	48,5	273	77	268	128,5
89		190	49	196	77	191	128,5
129	23,5	272	49	87	80	179	144
193	26,5	162	50,5	43	80,5	88	146,5
140	26,5	142	51	128	83	106	146,5
66	27	271	52	164	84,5	195	152
270	27	70	52	141	85	32	153
53	28	205	55	178	86,5	216	154
65	28,5	198	55,5	63	86,5	71	154,5
38	30	102	55,5	93	88,5		
160	31	131	57	37	89	104	157
163	31	40	59,5	42	89,5	137	165,5
33	31	100	60	46	89,5	67	180
95	32,5	183	60,5	96	91,5	45	189,5
136	33,5	43	61,5	185	91,5	115	201
197	34	56	61,5	182	95,5	44	215,5
126	34	58	62,5	62	96	122	225
221	34,5	79	62,5	105	98	159	231
97	37	139	64,5	258	100	85	231
107	37	101	65	68	100	83, 103	237
178	37,5	39	65,5	177	101,5	60	262
201	37,5	54	65,5	74	108		
173	38,5	269	65,5	92	111	75, 82	274
73	40	175	67,5	180	111	86	314
203	40,5	266	69,5	59	115,5	61	368
91	40,5	113	71	48	117	41	379,5
265	40,5	135	72	174	120	170	397,5
49	41,5	166	74	99	120	46	451,5
161	43	125	75	192	120	76	504,5
202	43	200	75,5	90	123	33	577,5
98	43	181	75,5	64	129	108	1351,5

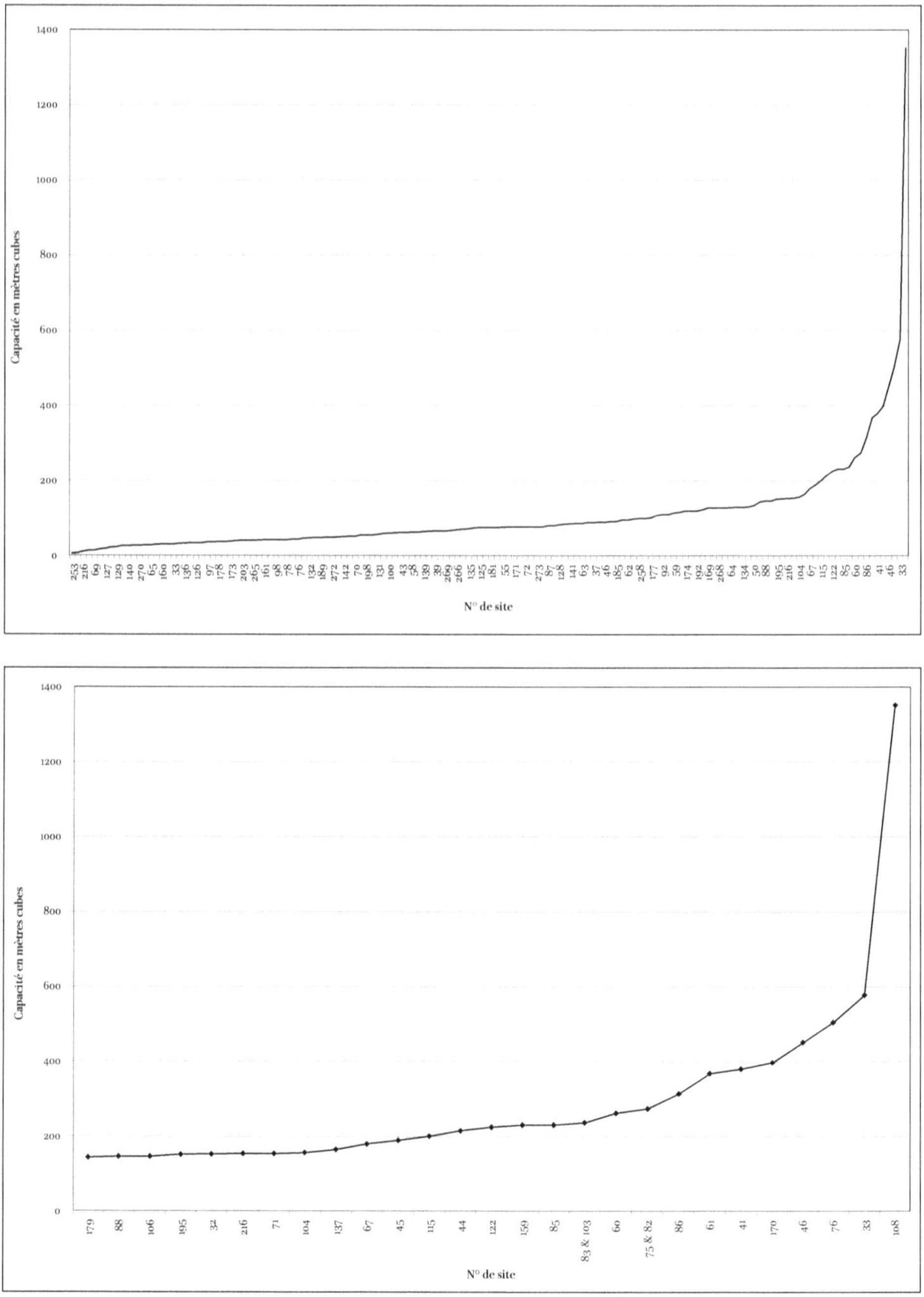

FIGURE 5.4 Graphe ordonné correspondant au tabl. 5.1, par capacité, dans l'ordre croissant ; a) ensemble du corpus du tabl. 5.1 (les n° de sites ne sont qu'indicatifs); b) sites présentant des capacités de stockage supérieure à 140 m3
A. FERDIÈRE

– à plus de 135 m^3, avec le site cat. 179 (27 sites),
– à plus de 160 m^2, avec le site cat. 137 (19 sites),
– à plus de 240 m^2, avec le site cat. 60 (10 sites) (cf. Fig. 5.4b),
– à plus 320 m^3, avec le site cat. 61 (6 sites),
– enfin à plus de 380 m^3, avec le site cat. 46 (4 sites).

Et l'on y note donc, avec ces paliers quantitatifs :
– seulement 2 sites[30] avec des capacités de moins de 10 m^3,
– 45 sites de 10 à 50 m^3,
– 53 sites de 50 à 100 m^3,
– 32 sites de 100 à 200 m^3,
– 12 sites de 200 à 400 m^3,
– seulement 3 sites[31] de 400 à 600 m^3,
– et enfin 1 seul site, vraiment exceptionnel, isolé, de 1351,5 m^3 (Saint-Aubin[32]).

On constate donc qu'une majorité (85 sites sur 148, soit 57,5 %) se situe entre 50 et 200 m^3, mais avec un lot non négligeable de sites ayant des capacités "moyennes" (de 10 à 50 m^3 : 30,5 %), les capacités les plus faibles comme les plus fortes restant plutôt exceptionnelles.

Il conviendrait de s'interroger ici sur la nature et les spécificités des quatre sites présentant les plus fortes capacités de stockage (plus de 400 m^3, dont 1 de 840 m^3), selon les critères proposés, à savoir, dans l'ordre croissant, Echternach I (cat. 46, Fig. 5.5), Seclin (cat. 76, Fig. 5.6, f et g), Voerendaal (pér. 1 ; cat. 33, Fig. 5.7) et donc, hors normes, Saint-Aubin (cat. 108, Fig. 5.8). Il s'agit en tout cas de grandes, voire très grandes *villae*.

En outre, le site de Voerendaal nécessite sans doute un commentaire particulier[33] : en l'absence d'une publication monographique, on ne peut s'appuyer que sur quelques articles succincts, la plupart en néerlandais[34]. L'auteur premier ne reconnaît d'ailleurs comme grenier (*horreum*) que le grand bâtiment attenant à la partie résidentielle, de type "à murets parallèles", de 375 m^2 de

30 Hardivilliers et Combs-la-Ville.
31 Echternach I, Seclin (pér. III) et Voerendaal (pér. 1).
32 Calcul de dimensions à partir du plan de prospection publié par Luc Jaccottey et repris ici grâce à lui (Fig. 5.8), avec mes remerciements, et non de celui figuré dans Ferdière et al. 2010, pl. 18, n° 1–22, dont l'échelle et donc les dimensions étaient erronées : ce réexamen en fait un cas absolument unique et hors normes, qui interroge…
33 Que l'on ne peut faire pour Saint-Aubin, en l'absence de fouille, le site n'étant connu que par les prospections aériennes contrôlées au sol.
34 Je remercie ici Raymond Brulet pour avoir bien voulu traduire pour moi certains passages ; on peut toutefois se référer, notamment pour les données carpologiques, à la thèse, en anglais, de Kooistra 1996, 129–252.

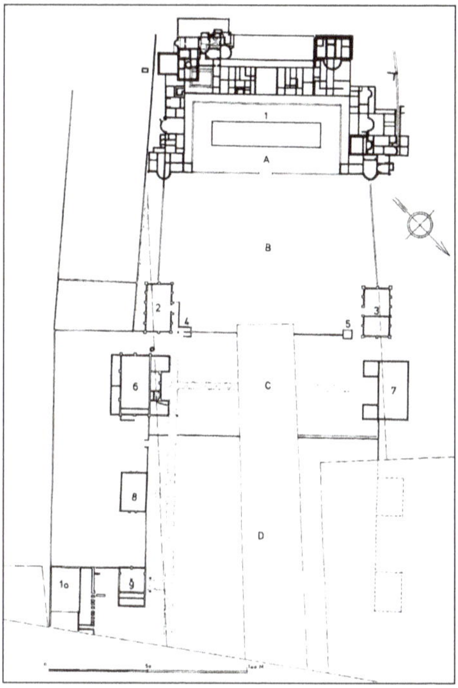

FIGURE 5.5
La *villa* d'Echternach (LU), avec ses greniers à contreforts
VAN OSSEL 1992, FIG. 145, D'APRÈS METZLER, ZIMMER, ET BAKKER 1981

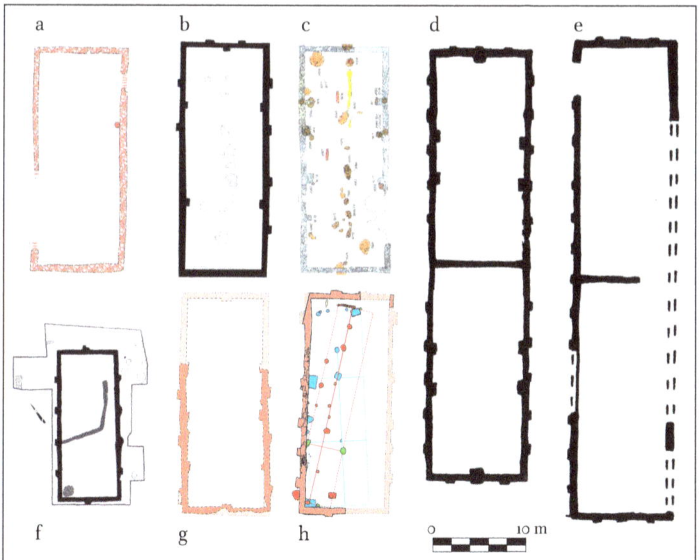

a. Loison-sous-Lens
b. Mériginies "Golf de la Pévèle", tranche 2
c. Hamois "Champion", bâtiment C
d. Seclin "les Hauts de Clauwiers", bâtiment P'
e. Seclin "les Hauts de Clauwiers", bâtiment Y
f. Oudenaarde "Mater"
g. Cysoing "Clos de l'Abbaye", bâtiment A
h. Cysoing "Clos de l'Abbaye", bâtiment B

FIGURE 5.6 Plans des deux grands greniers de Seclin (Nord), comparés à ceux d'autres établissements ruraux de Gaule du Nord
CENSIER 2013, FIG. 46, AVEC MODIFICATIONS

superficie utile, qui pourrait être des IIIe-IVe s. (?)[35]. J'ai jugé ici pertinent d'y ajouter comme greniers cinq autres bâtiments de la *villa* (quatre à contreforts[36] et un possible tour-silo), pour une capacité totale proposée de 746,5 m³, mais la question reste de savoir s'ils ont tous, à un moment, fonctionné ensemble[37].

Mais, clairement (Fig. 5.4b), la rupture principale de la courbe, après cette progression lente et régulière pour les sites à capacité de stockage les plus courants, se situe juste au-dessus de 260 m³, isolant ainsi seulement 10 sites aux capacités de stockage plus importantes – dont bien sûr, au sommet, les quatre sites dont il vient d'être question – : ce sont ensuite, par ordre décroissant, les sites cat. 41 (Cologne "Müngersdorf"), 61 (Marboué-Thuy), 86 (Berchères-les-Pierres), 75/82 (Saint-Gérard), 60 (Lürken), 170 (Granges), voir Fig. 5.9a à f[38], répartis dans des régions aussi diverses que la Beauce[39], la Belgique actuelle, le Centre-Est de la Gaule, les Germanies…

D'un point de vue chronologique, même si la très grande majorité des sites considérés concernent le Haut-Empire, ne peut-on constater une certaine réduction des capacités de stockage au Bas-Empire[40] ? Seulement huit greniers au maximum (sites cat. 33(?), 36, 41, 52, 99, 75/82, 178(?) et 216) sont possiblement datables du Bas-Empire (fin IIIe–fin Ve s. de n. è.). Parmi eux, les sites sur l'évolution desquels il est possible de raisonner à ce titre sont trop peu nombreux pour permettre de tirer des conclusions générales, mais l'on observe au moins deux cas, avec Echternach et Habay-la-Vieille[41], où l'édification tardive de tours-silos semble traduire une telle réduction de la capacité de stockage…

---

35 Période 3, selon Kooistra 1996, 130–35, Fig. 24c ; Willems (1987a, 49–50) estime pour ce grand grenier une capacité en grain de 300 à 400 m³, que je n'évalue qu'à 152 m³.

36 Dont le bât. A, au SE, où Willem J.H. Willems (1992, 528–29) reconnaît cependant le "travail des récoltes", compte tenu de la présence ici, à la fouille, d'une importante quantité de grain brûlé.

37 Pour ces points, voir *supra* ; ce semble être le cas pour la période 3, selon Kooistra 1996, *loc. cit.*

38 Auxquels on pourrait ajouter, en mettant la barre un peu plus bas, deux sites avec une capacité de 231 m³ (cat. 159, Bazoches-en-Dunois et 85, Biberist) et un de 237 m³ (cat. 83/103, Seeb), soit encore un en Beauce et 2 en Suisse (Germanie Supérieure).

39 Avec 2 sites, cat. 61 et 86.

40 Sur les greniers fortifiés du Bas-Empire des provinces germaniques, qui seraient à l'origine de tous les *burgi*, voir Bechert 1978 et Van Ossel 1992, 158–59 ; et voir par exemple le cas de Rheinfelden-Augarten : Asal 2005.

41 Et peut-être Voerendaal, cependant pris en compte ici pour la capacité globale de stockage du site (pour seulement 18 m³) ; pour ce site, le bâtiment A, que j'interprète aussi comme grenier, semble avoir continué à fonctionner au Bas-Empire (Van Ossel 1992, 365) : c'est le seul subsistant à la période 4 (IIIe/IVe au VIIe s.) du site, selon Kooistra 1996, 130–35, Fig. 24d.

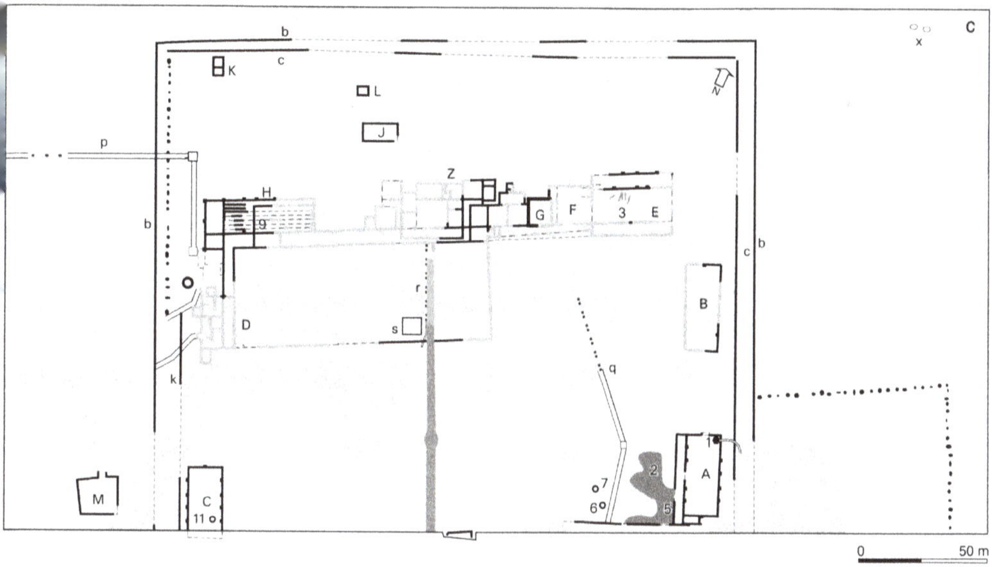

FIGURE 5.7 La *villa* de Voerendaal (NL) avec son ensemble de greniers
KOOISTRA 1996, FIG. 24C

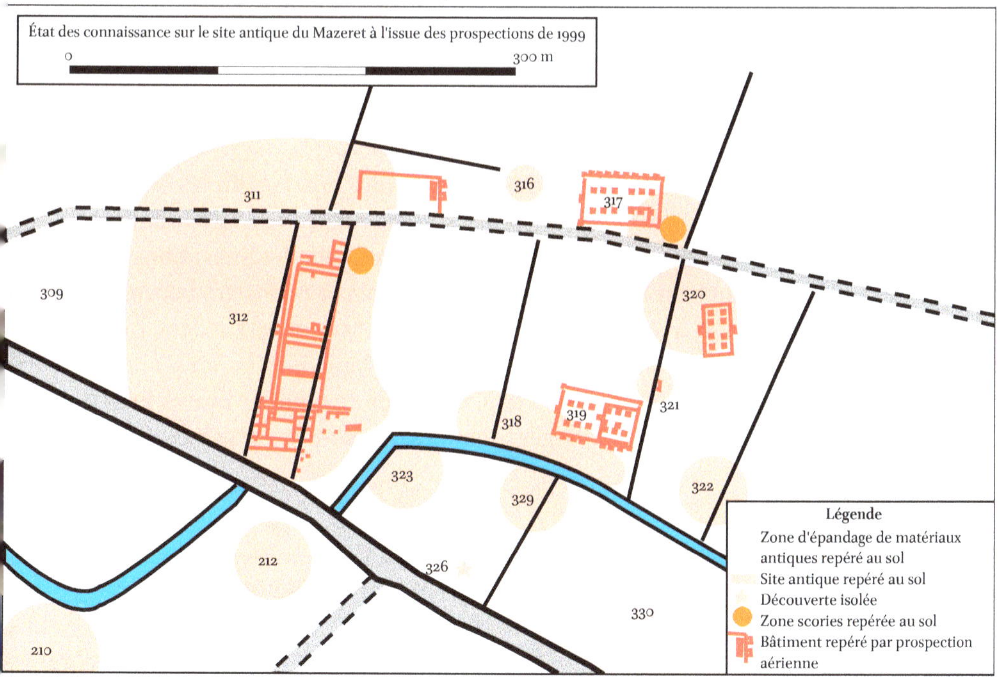

FIGURE 5.8 La *villa* de Saint-Aubin (Jura), avec ces trois grands greniers
DAO L. JACCOTTEY, D'APRÈS PHOTOGRAPHIES AÉRIENNES DE G. CHOUQUER ET PROSPECTIONS AU SOL DE L. JACCOTTEY

On pourrait en outre poser la question de l'éventuelle standardisation de greniers ruraux au Haut-Empire, quant à leurs dimensions : on raisonnera alors sur la surface (intérieure, ici prise en compte), pour un seul niveau de stockage. On prend dès lors en considération 188 greniers (pour 148 sites) pour lesquels cette donnée est connue avec une relative précision. On constate ainsi la constitution de groupes assez nets aux superficies équivalentes, parmi lesquels les plus pertinents (Tabl. 5.2), car les mieux dotés en nombre d'exemplaires et par ailleurs assez groupés, peuvent être situés entre 167 et 201,5 m², soit 33 items (17,5 % des greniers concernés)[42], ce qui révèle une forte représentativité.

Enfin, il faut malheureusement constater qu'il n'est pas possible d'aller plus loin quant aux modes de stockage des céréales (en grain battu et décortiqué, en épillets, en épis, en gerbes…) ou quant aux espèces de céréales concernées, car seuls deux des greniers inventoriés ici ont fourni des lots de graines brûlées ayant fait l'objet d'une étude carpologique : 2e état du grenier d'Alle (cat. 221, Fig. 5.2 ; orge et avoine majoritaire : Demarez et al. 2010, 306) et bâtiment A de Voerendaal (cat. 33, Fig. 5.7, avec surtout de l'orge) ; on pourrait à la rigueur y ajouter Biberist (cf. *infra* et Fig. 5.10 ; Blöck 2011-2012), avec les restes carpologiques cependant recueillis épars et non strictement dans le grenier retenu ici.

3 Les textes

À ce stade, sans doute est-il nécessaire de reprendre rapidement les données des textes antiques concernant les greniers privés, écrits des "agronomes" et autres auteurs latins ; ceci afin d'y rechercher notamment des indications chiffrées sur les volumes ou les épaisseurs de stockage ; et ce malgré leur relatif exotisme par rapport à nos contrées[43], en examinant la question dans l'ordre chronologique[44] :

1. Caton (IIe s. av. n. è. ; *Agr.* 92, chap. CI) ne donne que de brèves indications sur la manière de protéger le grenier (*granarium*) des charançons.

42 Et par exemple non moins de 13 exemplaires de 200 m², sans qu'on constate de standardisation régionale particulière, si ce n'est peut-être pour la cité des Ubiens (Kölstein), avec 2 des 3 greniers à 35–36 m² et surtout 4 des 14 à 200–201,5 m². Mais cette cité de Germanie Inférieure est la mieux documentée ici quant à ses greniers, ce qui fausse cette statistique en sa faveur.

43 À noter une possible mention de silos en Byzacène, dans *Afr.* 65 : […] *et omnibus fere villis sub terra specus condendi frumenti* […] (cavités souterraines pour contenir du grain).

44 Dont vérification faite dans le *Thesaurus Linguae Latinae*, *s. v. granarium* et *horreum*.

2. Chez Varron (milieu du Ier s. av. n. è. ; *R.* 1.57)[45], Stolon parle de greniers élevés, bien aérés et aux murs enduits.
3. Virgile (dernier tiers du Ier s. av. n. è.) n'évoque pratiquement pas le stockage/conservation du grain, si ce n'est pour dire qu'une bonne terre "fait craquer les greniers (*horrea*) sous des récoltes abondantes" (*G.* 1.49 ; 2.518).
4. Columelle (Ier s. de n. è. ; Col. 1.6.10 ; cf. aussi 12.2.2) mentionne les *granaria* bien aérés par des fenêtres, éventuellement voûtés, au sols en terre battue ou en *opus signinum*, aux murs enduits, et divisés en plusieurs compartiments pour les différentes espèces ; greniers aériens que l'auteur préfère, pour les contrées italiennes plutôt sèches, aux silos enterrés utilisés dans d'autres régions ; il indique en outre que le charançon n'affecte le blé que sur une paume (env. 7,3 cm) de profondeur et qu'il ne faut donc pas alors remuer le grain, pour éviter sa propagation (une information reprise dans Plin., *Nat.* 18.302).
5. Pline l'Ancien (3e quart du Ier s. de n. è. ; *Nat.* 18.301–308) consacre le chapitre LXXIII de son livre XVIII à la conservation du grain : confiné, sans air, dans des greniers (*horrea*, *granaria*) aux murs épais, de brique, ou en bois, alors au contraire bien aérés et surélevés sur des piliers ; avec des planchers suspendus ou non, revêtus ou non de *tegulae* ; il recommande d'engranger en épis et vante les avantages du silo enterré utilisé dans certaines régions de l'Empire ou extérieures, tout ceci selon la destination des récoltes. Parmi les contrées évoquées par cet auteur et le précédent à ce sujet (silos), malheureusement pour nous, aucune mention n'est faite des provinces gauloises.
6. Enfin, Palladius (au Ve s. de n. è., d'origine espagnole ; Pallad. 1.19) consacre un petit chapitre aux greniers, en n'évoquant que les *horrea* bien aérés, aux sols revêtus de brique, aux murs enduits et munis de compartiments pour les diverses espèces, reprenant alors Columelle quant aux charançons ; il mentionne par ailleurs (2.20.5) la possibilité d'engranger le blé en épis avec la tige, en attendant qu'il soit sec pour le battre.

Il est donc notable qu'aucun de ces auteurs antiques ne mentionne une épaisseur de stockage du grain optimale pour les greniers.

Ajoutons à titre indicatif les préceptes d'Olivier de Serres, au tournant des XVIe et XVIIe siècles, qui consacre le chapitre VI de son "*Second lieu*" (sur le grain) aux "*Greniers à blé. De la garde des Blés et de leur débite*" (Serres 1600,

45 Pour les *horrea*, à prévoir vastes si l'on privilégie la culture des céréales : Var., *R.*1.11.2 ; et voir aussi Var. *L.*5.106 ; et pour les *granaria* : *Men.* 185 ; et dans les *villae* : *R.* 3.2.6.

FIGURE 5.9 Les sites ruraux à greniers de 260 à 400 m^3 de capacité : a) Granges (Saône-et-Loire, FR ; DAO Pierre Nouvel, d'après Nouvel 2016, 4, site n° 251, plan p. 286, avec modifications) ; b) Cologne “Müngersdorf” (DE ; d'après Fremersdorf 1933, Taf. 13) ; c) Marboué “Thuy” (Eure-et-Loir, FR ; DAO A. Lelong, avec modifications) ; d) Berchères-les-Pierres (Eure-et-Loir, FR ; DAO S. Martin d'après un dessin original d'A. Lelong) ; e) Saint-Gérard (BE ; d'après Brulet 2008, fig. 490-491) ; f) Lürken (DE ; d'après Heimberg 2002–2003, fig. 42 et 46)

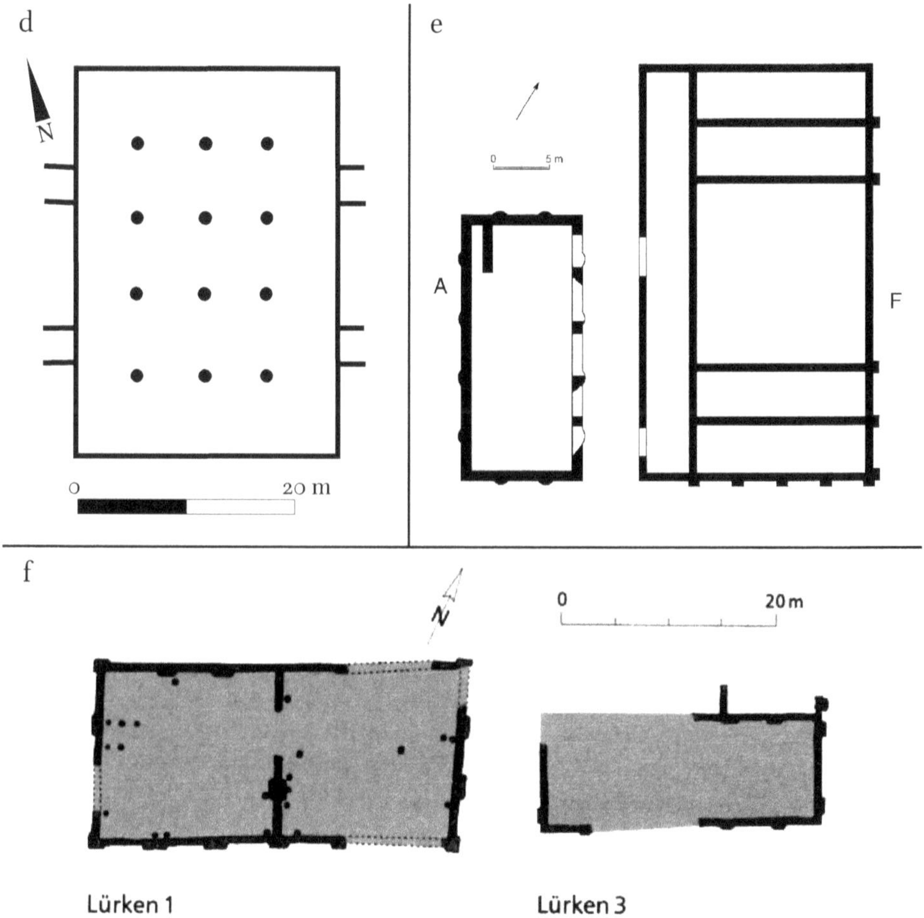

FIGURE 5.9 (cont.)

210–18) : après avoir proposé, au chapitre précédent (ibid., 208–09), de plus souvent adopter aussi dans ces contrées méridionales (qui sont en général celles concernées par son travail) la pratique septentrionale de la conservation en gerbes, avant battage, il préconise – pour les régions méridionales qui l'occupent – un grenier à l'étage de la maison rustique, ou bien en tout cas sur plancher surélevé "de deux bons pieds" (env. 70 cm ; il ajoute de garnir l'espace du sous-sol, pour l'assainir, soit de charbon de bois soit de scories de forge), en espace aéré au moins de deux côtés par des "fenestres" qui puissent être ouvertes ou fermées à volonté. S'il est situé à l'étage, le plancher d'aix (lattes) jointifs doit être "quarrelé de briques", avec des aires de stockage assez vastes pour remuer le grain aisément, dans des caissons (parois de bois sur les côtés des

TABLEAU 5.2 Tableau des greniers par groupes de superficie interne (sur un seul niveau), en m² : nombre d'exemplaire par groupe ; en gris, les groupes les mieux représentés

Surface int. grenier (en m²)	N. d'ex (si ≥ 3)
35–36	3
69–70	5
81	3
87,5–88	3
104–106	5
105	3
112–112,5	5
120	3
126–128	8
126	5
128	3
167–171	7
175–176	5
194–198	7
196	3
200–201,5	14
200	13
312–312,5	3
360	4

FIGURE 5.10 Coupe restituée du grand grenier axial (bât. D) de la *villa* de Biberist (CH)
SCHUCANY 2006, FIG. 10/D15

tas) distincts selon les espèces ; les céréales doivent en tout cas y être engrangées bien sèches et si possible donc bien séparés par espèce ; en cas d'infestation, un bon pied (env. 35 cm, soit plus que ce que recommande Columelle) doit être ôté du dessus du tas (dont il ne précise toutefois pas quelle est l'épaisseur de stockage optimale, en tout cas donc ici supérieur aux 35 cm de notre " modèle") ; il évoque ensuite le mode de conservation, curieux à ses yeux, en fosses excavées, vastes et profondes, pratiqué alors en Guyenne et Gascogne (les *cros*), en rappelant à ce propos les textes de Pline et Varron au sujet de ces silos ; il précise justement enfin que les durées et facilités de conservation du grain varient selon les espèces ; les prélèvements au grenier se feront ensuite au rythme des besoins divers, d'où l'importance de bien y entretenir le grain, en en remuant les tas à la pelle "trois ou quatre fois de l'an"... Ces prescriptions complètent dont utilement, certes bien plus tard, celles des agronomes latins.

Par ailleurs, les textes antiques mentionnant strictement des greniers en Gaule sont peu nombreux et à ma connaissance tous tardifs, à l'exception de deux premières références, allusives de stockages ou transports de blé des Gaules au Ier s. de n. è. (pour les importations de blés des Gaules à Rome, voir récemment Bonacci 2015, qui ne mentionne pas les deux textes qui suivent) :

a. Tacite (*Hist.* 5.23) mentionne le transport de blé par voie d'eau de Gaule pour le blocus final de Civilis, en 70 de n. è., pour la Germanie Inférieure.
b. Pline l'Ancien, quant à lui (*Nat.* 18.66), dans le 3e tiers du Ier s. de n. è., évoque l'importation de blés "légers" gaulois à Rome.
c. Le *Panégyrique de Maximien* par Mamertin (du 21 juillet (?) 291 : *Paneg.* III(11).15.4), avec une mention concernant sans doute la Gaule de l'est et peut-être la cité des Trévires, de moissons qui, grâce à Maximien, "font ployer les greniers (*horrea*) où elles sont rentrées", et du fait que "du reste, les terres cultivées ont doublé d'étendue", des terres ensemencées ayant pris la place de forêts.
d. Ammien Marcellin (IVe s. ; Amm. 10.3 et 5) mentionne des greniers sur la Saône.
e. Quant à Ausone, au IVe s., il note que, pour son "petit héritage" de *Pauliniacum*, dans le Bazadais (*Hered.* 17–28), avec ses 200 jugères de terres de labour[46], il engrange toujours deux années de récolte d'avance, sans doute pour deux années de consommation ; mais, une année à court de blé (*Ep.* 20), il en fait venir du domaine proche de son ami Paulin de Nole à *Hebromagus*.

46 1050 jugères au total, soit 262,5 ha, pour les 50 ha labourés desquels P. Ouzoulias (2006, 222) propose, avec une jachère annuelle par moitié (soit alors 25 ha), une récolte annuelle de blé d'au moins 187 hl.

f. Notons encore les mentions, à la toute fin du IVe s., de greniers sur le Rhône et des blés gaulois – dont des Sénons ou des Lingons – approvisionnant l'Italie (Claud., *Carm.* 18.404–405 ; *Carm.* 22.392–396 ; *Carm.* 24.91–98).

g. Quant à Sidoine Apollinaire, au Ve s., on note le domaine privé de *Martialis/Vialoscensis* de Maurisius (Sidon., *Epist.* 2.14.1, vers 465 de n. è.), avec ses *horrea* (traduit "greniers"), qui aurait servi de quartiers d'hiver aux légions de César (peut-être Marsat, Puy-de-Dôme ?) ; ou les *horrea* des bords de Saône et du Rhône que l'évêque de Lyon Patiens a fait remplir de blé (*frumenta*) à partir de ses propres greniers (Sidon., *Epist.* 6.12.5–8, en 471–472) ; ou encore pour son domaine de Bourg-sur-Gironde (Sidon., *Carm.* 22.169–170), avec ses *horrea* aux bâtiments étendus en longueur ; enfin, incidemment, pour la richesse en blé de la Limagne d'Auvergne (Sidon., *Carm.* 7.139–152 ; *Epist.* 4.21.5).

h. Enfin, Grégoire de Tours, au VIe s. (mais mentionnant souvent des événements antérieurs, remontant aux IVe et Ve s.), évoque parfois des greniers dans son *Histoire des Francs*, mais seulement sous le vocable *horreum*[47] : pour la résistance de l'évêque Injurius au roi Clothaire pour le prélèvement fiscal à l'Église (*ab eorum stipe tua horrea repleantur* : *Franc.* 4.2, qui semblerait indiquer un stockage en gerbe dans les greniers du roi) ; à peu près à la même époque (VIe s.), dans un monastère non indiqué, les moines sortent de leur greniers du grain pour le faire sécher au soleil (*de horreo anonas* [...] *ponerent.* [...] *in horrea* [...] *triticum...* : *Franc.* 4.34) ; lors d'une épidémie sous Chipéric, dans une prière, Frédégonde mentionne les greniers pleins (*horrea replebantur frumento* : *Franc.* 5.35)[48].

On pourrait ajouter ici, en ne se fondant que sur le *Thesaurus Linguae Latinae* (*s. v.* "*horreum*"), que, pour les inscriptions au *Corpus Inscriptionum Latinarum*, on note une absence de mentions épigraphiques pour les Gaules et seulement trois occurrences concernant les Germanies : CIL, XIII, 7441, 7749 et 11540.

---

47 Et ni *granarium* ni *spicarium* ; recherche faite à mon attention en ligne par E. Zadora-Rio, que je remercie ici, sur le site des *Monumenta Germaniae Historica* de Brepols.

48 Deux autres mentions d'*horrea* dans l'*Histoire des Francs* ne concernent pas la Gaule mais respectivement Babylone et Rome.

4 L'épineuse question des rendements

En définitive, peut-on, à partir de ces réflexions sur la capacité des greniers, déduire, de l'apparente augmentation de capacité de stockage en céréales des établissements ruraux entre la période laténienne et celle romaine en Gaule, un accroissement massif de la production agricole ?

Sans bien sûr nier l'évidence de cet accroissement, avec la mise en culture massive de nouvelles terres et l'augmentation des rendements que l'on observe semble-t-il alors, sans doute faut-il cependant reconnaître une certaine tendance à la surévaluation, d'une part pour les capacités de stockage avec ces greniers massifs, d'autre part par conséquent pour la production céréalière gallo-romaine.

Cette vision peut-être trop optimiste de la production céréalière en Gaule romaine peut avoir pour cause un effet de sources des données archéologiques elles-mêmes, car les grands greniers maçonnés gallo-romains sont quand même plus aisément perceptibles que leurs équivalents en bois (il est vrai en général de plus petites dimensions) de la période antérieure, et d'ailleurs aussi postérieure (Haut Moyen Âge).

Et l'évaluation est sans doute également faussée par – justement – les présupposés concernant la période romaine, avec l'impression d'accroissement massif de la production agropastorale à cette époque : encore faudrait-il s'assurer – à ce stade – que cette vision n'est pas, précisément, justement due à l'observation de ces vastes greniers en pierre, ce qui nous ramènerait en quelque sorte à un raisonnement en boucle… En effet, la production céréalière de la Gaule indépendante (cf. *infra*) était déjà loin d'être négligeable puisqu'elle apparaît sans doute comme le principal objet de convoitise de ces territoires de la part de Rome, qui aurait motivé notamment la conquête césarienne.

Mais il est nécessaire de faire aussi entrer en ligne de compte dans ce débat une autre considération : celle, globalement, de la dimension économique, avec le passage – certes progressif –, d'une période à l'autre, d'une économie plutôt domestique, autosuffisante, à un système qui doit être qualifié d'économie de marché à la période impériale, affectant profondément les provinces (circulation des biens de consommation, approvisionnement des villes ainsi que de l'armée…). Dès lors, même pour une production à peu près équivalente entre les deux périodes, les nécessités de stockages deviennent différentes, une partie parfois importante de la production de chaque exploitation étant non plus seulement destinée à l'autosubsistance des protagonistes ruraux, ainsi qu'au renouvellement des semences, mais à l'exportation plus ou moins lointaine, vers les agglomérations et chefs-lieux de cités, voire Rome, l'Italie et l'armée (ici, *Limes* rhénan). Ceci n'a pu que jouer un rôle déterminant dans la modification des modalités et des capacités de conservation et stockage des céréales…

Il n'en reste pas moins que les raisons de stocker et/ou conserver à plus ou moins long terme des quantités importantes de céréales restent en principe multiples :

a. la réserve de semence (en préservant le pouvoir germinatif des grains) pour l'automne suivant, voire pour certaines espèces et/ou pratiques, pour semis de printemps, donc à plutôt court terme ;
b. le stock nécessaire à l'alimentation même des paysans et de leur famille, ou de la *familia* et des servants de la *villa* et de l'exploitation pour une année, soit des dizaines, voire centaines de personnes à nourrir pour les plus grands établissements ;
c. une réserve de sécurité en cas de récolte désastreuse ou nulle, au moins pour une année, tant pour le semis que pour l'alimentation *in situ* ;
d. les stocks destinés à l'extérieur à plus ou moins long terme : prélèvement fiscal en nature, négoce de grain, à destination des villes et autres agglomérations non agricoles, ainsi que de l'administration et surtout de l'armée, et donc pour une exportation plus ou moins lointaine, enfin.

Quant à ce dernier point, il est très difficile, voire impossible – du moins à partir des seules sources disponibles en général pour nos contrées, les données archéologiques – de mesurer les proportions destinées d'une part au prélèvement obligatoire, de l'autre au négoce. Les sources écrites ou épigraphiques sont peu nombreuses pour les Trois Gaules et Germanies, si ce n'est par exemple, au Bas-Empire, quelques mentions attestées de greniers (publics) sur les bords de la Saône (cf. *supra*).

Quelle est donc alors, entre ces quatre raisons principales du stockage, la finalité fonctionnelle principale de ces imposants greniers observés sur les établissements agropastoraux de Gaule du Nord ? Les deux premières (a et b), compte tenu des capacités concernées ici, semblent devoir être exclues. On peut penser que l'on a bien affaire ici, essentiellement, à un stockage temporaire (quelques mois ?) destiné à l'extérieur (d), qu'il s'agisse de prélèvements obligatoires, imposés pour au moins une part ou – sans doute surtout – de commerce. On se place bien dès lors dans le cadre d'une économie de marché dans laquelle le commerce des céréales, à destination des villes, de l'armée ou d'exportations plus lointaines, occupe une place essentielle.

Ces changements dans les capacités de stockage rural depuis la période laténienne (cf. *infra*) sont donc à mon sens surtout révélateurs d'une modification assez fondamentale, d'ordre économique, dans les relations entre production alimentaire et consommation, dans les systèmes d'échanges, dans le cadre maintenant de cette économie de marché, ou en tout cas plus ouverte.

Mais on ne peut nier (cf. *supra*) que ces grands greniers sont sans doute aussi – dans une certaine mesure – révélateurs d'une augmentation de la production céréalière, avec plus de terres cultivées mais aussi sans doute des rendements accrus. Ces questions, ainsi que celle, corrélée, du réel niveau technique de l'agriculture romaine, sont complexes et délicates : Pierre Ouzoulias s'y est essayé, avec quelque succès, dans sa thèse (Ouzoulias 2006, 31–33, 75–76, 98–104, 171–87, 208–14…), son apport principal étant historiographique, en revisitant les auteurs "classiques" sur le sujet, tels Adolphe Dureau de La Malle (1840) ou Johan K. Rodbertus (1907), ainsi que des sites emblématiques tels que Chiragan, pour lequel Léon Joulin (1902) proposait, sur des bases de calcul éminemment discutable, des rendements de l'ordre de 12,5 hl à l'hectare… (voir aussi ma discussion à ce sujet avec P. Ouzoulias *in* Ferdière 2009, 222-23, notice 219.)

En résumé (et donc essentiellement pour des régions méditerranéennes), Varron (*R.* 1.44.2) et Pline l'Ancien (*Nat.* 18.94–95) – avec des rendements exceptionnels de 100 grains récoltés pour 1 semé (en Sicile, Bétique et Égypte) voire 150 pour 1 (en Byzacène) et une production de 10 *modii* par jugère[49] (en céréales vêtues, et surtout sans doute en blé[50], "*triticum*") – sont sans doute dans l'exagération, à moins qu'il ne s'agisse de semis en poquets. Mais le premier mentionne de rendements moyens de 10 pour 1, et 15 pour 1 en Étrurie ; la mention de Columelle (3.3.4), avec 4 pour 1, soit 9 *modii* de semence (bonnes terres) ou 10 (moins bonnes terres) pour 3,1 à 3,4 hl de rendement à l'hectare, se rapporte selon lui à l'ancien temps (soit sans doute la période républicaine), et paraît toutefois plus raisonnable, au regard des rendements de l'agriculture préindustrielle de nos contrées.

On peut noter d'autres estimations :

a. pour Alle (Suisse, cat. 125/221, cf. Fig. 5.2), les auteurs ont tenté une estimation de la surface du domaine exploitée en orge[51], pour le grenier mis au jour contenant un poids de grain évalué à 23 720 kg pour une surface de stockage proposée de 240 m² de grain, 4 ou 5 fois supérieure aux capacités de stockage des phases 1 et 2 du site (Demarez et al. 2010, 393-94) ;
b. pour la *villa* de Champion à Hamois (BE, cat. 95/47, Fig. 5.6, c ; Van Ossel et Defgnée 2001, 232), les auteurs, retenant une surface de stockage de

49 1 *modius* = 8,788 litres ; 1 jugère = 25 ares.
50 "Blé poulard" selon la traduction du texte de l'édition utilisée ici (J. Heurgon) pour Varron.
51 La céréale la mieux représentée dans l'étude carpologique réalisée pour ce site.

36 m², propose un stockage théorique de 10 tonnes pour une épaisseur de 0,40 m maximum ;

c. quant à la *villa* de Voerendaal (*supra* ; cat. 33, Fig. 5.7), l'auteur (Willems 1987a, 49-50) estime, pour le seul grand *horreum* à murets parallèle dont il évaluait la capacité à au moins 300 m³, soit environ 235 000 kg, qu'il était nécessaire de disposer de 150 ha des fertiles lœss présents dans cette région ;

d. Lars Blöck propose quant à lui de distinguer, parmi ces greniers de grande capacité, ceux qui correspondent à la production d'un établissement et ceux qui pourraient recevoir les récoltes, centralisées, de plusieurs fermes (Blöck 2011-2012, 100–110) ; il isole même, comme alors vraiment public, le grand grenier de Walldorf (non pris en compte ici car pour moi atypique ; cat. 275 : Rabold 2001, 2005, 356–58) ; il met en œuvre les sites de Heitersheim, Biberist, Bad Rappenau (2 sites) et Voerendaal (respectivement cat. 254, 85, 251 et 33), qui ont tous quatre fourni des céréales carbonisées identifiées grâce aux analyses carpologiques, pour proposer des surfaces de domaines (Blöck 2011-2012, 92–109) ;

e. enfin, M. Poux (Poux 2017 ; voir aussi Poux et Borlenghi 2016) a récemment proposé, pour les greniers de très grande capacité du type Panossas (à murets parallèles internes, de soutien pour plancher surélevé : mon type 5), d'y voir des équipements sans doute publics destinés à l'approvisionnement de l'armée dès le Ier s. de n. è., avant donc la mise en place de la véritable annone militaire, à partir des Sévères ; il se réfère notamment au constat qu'apparemment, les régions rhénanes ne suffisaient pas alors (2e moitié du Ier s.) à l'approvisionnement de l'armée sur le *Limes* de Germanie Inférieure (notamment lors du blocus de Civilis en 70 : voir Reddé 2011b). Je ne m'aventurerai pas à aller jusque-là, mais ces points restent à discuter…

5 Quels changements entre La Tène finale et la période romaine ?

Pour revenir à La Tène finale, on ne connaît pas, en effet, pour cette période, de telles capacités de stockage de céréales, alors que, à partir de La Tène C2, les grandes batteries de silos antérieurement attestées semblent avoir pratiquement disparu (voir Chap. 4 dans le présent volume, auquel on se réfèrera pour plus de détails sur la situation protohistorique). Le site du "Clos des Primevères" à Entrammes (Mayenne) (Fig. 5.11), daté de La Tène C2/D1a, se distingue cependant, de manière pour l'instant assez unique pour cette période (Guillier

FIGURE 5.11 L'ensemble de greniers de La Tène finale d'Entrammes (Mayenne)
GUILLIER ET AL. 2015, FIG. 12

et al. 2015 ; Bossard 2015 ; Bossard et al. 2018[52]). Il pourrait s'agir non d'un établissement rural mais d'une agglomération fonctionnant juste avant l'*oppidum* du Port du Salut, comportant plus de 80 greniers, de petites dimensions et organisés en rangées parallèles, sans doute détruits par un incendie ; les prélèvements pour étude carpologique ont montré que le stockage concernait à 82 % du blé amidonnier, avec du millet, de l'orge vêtue, de l'épeautre, de l'avoine et diverses adventices ; les capacités de ce stockage dépasse évidemment les besoins du site pour concerner le commerce. Avec ce grand nombre de greniers surélevés très généralement à quatre poteaux, de petites dimensions (en moyenne 3 × 3 m, soit 9 m^2 par grenier), on atteindrait une superficie de stockage de plus de 720 m^2, c'est-à-dire 252 m^3 de capacité totale de stockage sur une épaisseur de 0,35 m. La multiplication des petites structures compenserait ici les petites dimensions, pour un stockage massif. On peut envisager que la séparation en multiples lots distincts et distants permette en principe de mieux prévenir la propagation d'incendies, pour des structures en bois sans doute couvertes de chaume.

On peut comparer ce site laténien[53] avec d'autres comportant de telles concentrations de structures de stockage, à Inzinzac-Lochrist (Morbihan), avec ses 46 greniers de La Tène ancienne, et à Courseulles-sur-Mer (Calvados) (Jahier 2011, 226), ou à Val-de-Reuil (Eure) (Moreau, Granier, et Dubreucq 2015). Un nombre assez important de greniers organisés en quartiers existent aussi sur les sites des "Natteries" (phase 3) ou de "la Bleure", à Cholet (Maguer et Lusson 2009, fig. 28-29). Et des sites de fermes dites aristocratiques de La Tène montrent aussi apparemment de fortes capacités de stockage de céréales, sans doute collectif : c'est le cas de Paule (Côtes-d'Armor), avec un vaste édifice rectangulaire, sur de puissants poteaux, dont une ligne centrale (38 × 10 m), pour la phase 4–5 (Menez 2009). Et l'on pourrait encore mentionner le grand bâtiment carré inédit, possible grenier, de Trégueux (Côtes-d'Armor), à parois rejetées et triple ligne de poteaux internes, de 52 × 52 m dont 36 × 36 m utiles au centre. De fortes capacités de stockage sont encore notées par exemple pour le site de La Tène Finale de Poulainville, près d'Amiens (Malrain et al. 2016).

52 Je remercie les auteurs d'avoir bien voulu me communiquer le texte de cet article avant publication.

53 Informations fournies par Patrick Maguer, François Malrain, Véronique Matterne, que je remercie ici ; complétées ici par Bossard et al. 2018.

En conclusion

Un saut quantitatif important, quant à la capacité de stockage des greniers ruraux, est franchi avec la mise en place du système provincial romain en Gaule. Ces changements survenus entre la période laténienne et l'époque romaine sont apparemment surtout imputables aux nouveaux objectifs attribués au stockage de céréales en contexte rural.

Il n'a en revanche pas été possible d'observer les évolutions en la matière à travers la période impériale en Gaule du Nord, et notamment entre Haut et Bas-Empire : les exemplaires de greniers tardifs attestés sont en effet trop peu nombreux, avec seulement sept sites pour les greniers à contreforts[54] et deux pour les greniers à piliers internes[55], auxquels s'ajoutent – ici – un site à grenier à murets parallèles[56] et au moins deux greniers-tours[57].

De nouveaux changements, profonds, se feront jour au Haut Moyen Âge.

---

54 Cat. 36, 05/47, 49, 41, 270, 76 et 33 (Anse, Emptinne, Fontaine-Valmont, Cologne "Müngersdorf", Saint-Quentin, Seclin pér. IV et Voerendaal pér. 2).

55 Cat. 178 et 99 (Mareuil-lès-Meaux et Obermendig).

56 Cat. 221, Alle, phase 4.

57 Cat. 46 et 216, Echternach II et Habay-la-Vieille.

CHAPTER 6

Storage in a Non-Villa Landscape: the Batavian Countryside

Stéphane Martin

The part of the Netherlands lying south of the Rhine corresponded roughly, in the Roman period, to the *civitas Batavorum* (Fig. 6.1).[1] This whole area came under Roman domination after the Gallic Wars in the 50s BCE. From around 20 BCE, large numbers of Roman soldiers were stationed in and around Nijmegen and, from the 40s CE onwards, along the Rhine down to the North Sea (see Polak and Kooistra 2013 for a historical and archaeological overview of the early Roman period). Both military and civilian sites have been the subject of extensive research; as elsewhere in Europe, rescue archaeology has considerably increased the available data. Furthermore, the wet climate and subsoil of the region have made possible a substantial number of palaeobotanical studies. The Batavian area thus appears particularly suited to a study of the provisioning of the Roman army and of its impact on the surrounding countryside.

It is remarkable that in spite of the heavy influx of military personnel and of the creation of Nijmegen/*Oppidum Batavorum* and other smaller urban centres that marked the beginning of urbanization in the region, the occupation pattern of the countryside did not differ much from the pattern known in the Iron Age (Willems and Enckevort 2009, 81–94, largely based on Willems 1986b and focused on the region around Nijmegen; Vossen 2007, 37–40, with a broader scope. Occupation ceased almost everywhere in the late third century: Heeren 2015). Only from the late first century CE onwards was there a moderate increase in the number of settlements. Most notably, there was very strong continuity in the architectural tradition; stone was extremely rare and *villae* were almost completely absent: this is the "non-villa landscape" described by N. Roymans (1996).

1 Originally, it probably extended from the territory of Nijmegen in the east, to the North Sea in the west; at a later date (mid or late first century CE?), the western part became the *civitas Cannenefatium* (on this complicated question, see Raepsaet and Raepsaet-Charlier 2013, 213–18; I leave out the problem of the *civitas Frisiavonum* since no site from this region is discussed here). Though the focus of this paper is on the eastern part of this area (Batavians *stricto sensu*), I will also take up some examples from the western part, as the countryside of both *civitates* displayed similar trends.

The continuing tradition of indigenous settlement and the lack of *villae* have been generally taken as a sign that crop husbandry, if not agrarian production on the whole, was still low in the Roman period, and that most of the foodstuff needed by the army and the city-dwellers was imported, consolidating the general impression of a peripheral region in the Roman empire. This impression is consistent with the lack of good agricultural land outside of the fertile riverine area: settlements were concentrated there and their numbers decreased markedly in peat and sandy areas (see van Dinter 2013, Appendix 1, and Polak and Kooistra 2013 for palaeogeographic maps of the region). The presence of cereals not grown in the region at various military sites, as well as the boat loaded with grain found in Woerden and dated by dendrochronology to the last quarter of the second century CE, also lend support to the idea of continuing imports throughout the Roman period (Kooistra 2009, 2012; on Woerden, Pals and Hakbijl 1992 [carpology]; Haalebos 1996 [archaeology and dendrochronology]).

However, this view has been challenged in recent years by the "Sustainable Frontier" project (see Polak and Kooistra 2013 for a general presentation). Research carried out in this framework showed that the carrying capacity of the region was higher than previously thought. Most notably, in a paper published in 2009 in the *Journal of Roman Archaeology* (*JRA*) and thus disseminated to a wide audience, Maaike Groot, Stijn Heeren, Laura Kooistra and Wouter Vos defended the idea that during the Mid Roman period (from the end of the first to the end of the third centuries CE), the Kromme Rijn region was indeed able to produce agrarian surpluses, both of animals and of cereals, in quantities not to be overestimated but still sufficient for peasants to participate substantially in a market economy (Groot et al. 2009; focused on animal husbandry, Groot 2016 sticks to this general framework).

In reaching its conclusions, the paper combines archaeological, archaeobotanical and archaeozoological evidence from two fully excavated and published sites from the Kromme Rijn area (around Utrecht), namely Tiel "Passewaaij" (cat. 312) and Wijk bij Duurstede "De Horden" (cat. 313), taken as representative of the overall situation. The existence of surpluses is argued for on qualitative and quantitative grounds. Qualitative arguments stem from archaeobotanical and -zoological data, showing specialization in animal husbandry but not in cereal cultivation. The quantitative assessment comes from the comparison of the protein requirements of the inhabitants, whose number is deduced from the number of houses per settlement and per phase, with the production capacity of each settlement, reconstructed from the size of stalls and granaries. Furthermore, some granaries also changed in shape, exhibiting more "Roman" features. The intensification in cereal production is therefore

CITÉS ROMAINES DES ESTUAIRES DU RHIN, DE LA MEUSE ET DE L'ESCAUT
Marie-Thérèse et Georges Raepsaet-Charlier
DAO Nathalie Bloch, © CReA-Patrimoine - 2013
· Frontière de province (provincia)
· Frontière de cité (civitas)
· Voies romaines
· Fleuves et rivières
· Chef-lieu de civitas
· Agglomération
· Site romain
· Sanctuaire
· Camp militaire
· Tracés modernes des îles et estuaires
N
0 20 km
Castricum
Assendelft
Velsen
Oer-IJ
Amsterc
Katwijk
Rijnsburg
Valkenburg
Leiden
Oude Rijn
Alphen
Wassenaar
Zoeterwoude
Zwammerd
Hazerswoude
Bodegraven
Scheveningen
's-Gravenhage-Den Haag
Scheveningenweg
Voorburg
Monster
's-Gravenzande
Delft
Ockenburg
Rijswijk
Moordrecht
Poeldijk
Naaldwijk
C. Cannanefatium
Helinium « Swarte Waal »
Oostvoorne
Brielle
Schiedam
Rotterdam
Vlaardingen
Nieuwe Maas
Lek
Rockanje
Poortugaal
Rhoon
Alblasserdam
Oude Wereld
VOORNE
Spijkenisse
Giessendam
GOEREE
Ouddorp
Ouwe Stee
Oude Oostdijk
Goedereede
Putten
Mijnsheerenland
HOEKSCHE WAARD
Puttershoek
Dordrecht
Dubbeldam
's Gravendeel
C. Frisiavonum
Maasdam
BIESBOSCH
Numansdorp
Strijen
Oude Maas
Strijensas
Burg
Haamstede
Duivendijke
OVERFLAKKEE
Westenschouwen
Brabers
SCHOUWEN-DUIVELAND
Roompot
Zierikzee
St. Philippaland
Striene
Ter Heyden
GERMANIA
Colijnsplaat
Oostkapelle
Vrouwen-Polder
NOORD-BEVELAND
THOLEN
Bred
Oranjezon
St. Maartensdijk
Poortvliet
Domburg
Westkapelle
Veere
Tholen
Ooster Schelde
WALCHEREN
Venusdam
Middelburg
Bergen op Zoom
Rijsbergen-Zundert
Aa/Weerijs
Mark
Koudekerke
C. Menapiorum
Reimerswaal
ZUID-BEVELAND
Vlissingen
Ritthem
Borssele
Breskens
Honte
Ellewoudsdijk
Hoogstraten
GALLIA BELGICA
Aardenburg
Maldegem
Ekeren
Kleine Schijn
Wijnegem
Grobbendonk

FIGURE 6.1
Map of the region discussed in this chapter
M.-T. AND G. RAEPSAET-CHARLIER, N. BLOCH, © CREA-PATRIMOINE, 2013

inferred mainly from this architectural change. According to the authors, these surpluses were directed to the new urban and military markets, and led to the intensification of the monetary economy, as witnessed by the numerous "imported" objects found in rural settlements.

This article thus alters our perception of the area and raises important questions. Together with other papers from the "Sustainable Frontier" project, it contributes to a more balanced view of the Batavian territory, which was less underdeveloped than previously assumed. Nonetheless, the idea that the area experienced a significant increase in crop cultivation and yielded enough surpluses to feed the producing and part of the non-producing population is at odds with the fact that Batavians cultivated the same range of crops from the Iron Age down to Late Antiquity/Early Middle Ages (when rye was introduced). This contrasts strongly with what can be observed in more southerly regions (see the works of Véronique Zech-Matterne for northern France) and raises questions as to whether the western Lower Rhine area actually ever experienced significant agrarian changes in the Roman period.

As the latest and most comprehensive attempt at quantifying agrarian production on the Lower Rhine, the *JRA* paper is the obvious starting point for new research on the subject. Assessing its results appears all the more necessary that they are now taken as hard data and serve as starting points for new research projects, rather than being discussed as working hypotheses (see e.g. Finding the limits of the Limes 2012). This will be the aim of the present chapter, building on the new method for calculating the storage capacities of granaries put forward earlier in this book (see Chapter 3 in this volume) and making use of new archaeological data from the region. Although the quality of the excavations themselves is of a high standard, they generally concern only part of the settlements. It is for this reason that the 2009 paper commented on above only used the evidence from two fully excavated sites: only then is it meaningful to attempt correlations between demographics, storage capacities, and sown areas. Since then, more extensive excavations have been published, most notably in Geldermalsen "Hondersgemet" (cat. 304) and Cuijk "De Nielt" (cat. 317). Furthermore, partial excavations can still yield interesting information, e.g. on the changes in the shape and size of granaries through time. Augmenting the sample is all the more important as granaries, and wooden structures in general, are not often precisely dated in the Netherlands. Due to the sandy nature of the soils, a clear stratigraphy is often lacking and material directly associated with buildings is rare. Since some building types, particularly outbuildings such as granaries, remained unchanged from the Bronze Age to the Middle Ages, it is not possible to date them on the basis of their plan alone. This is why in Oss "Ussen", a well-excavated settlement occupied from the eighth century BCE to the third century CE, 84.15% of the 479 granaries are not dated (Schinkel 1998, 255–66; here cat. 22/311). In Wijk bij Duurstede "De Horden", one of the two sites studied in Groot et al. 2009, 61.79% of the 89 granaries remain undated and only 26.96% of all granaries are dated to a single phase of occupation (Vos 2002, 111–12). In more recent publications, granaries tend to be distributed among occupation phases with greater confidence, on the basis of situation, orientation and vicinity to other structures (see for instance Tiel "Passewaaij": Heeren 2009, 38–48). But this is not always the case: in the excavation report from Cuijk "De Nielt", 39 of the 62 structures identified as granaries (62.9% of the total) are left without a date (Habermehl and Renswoude 2017, 854–71). In this paper, I have retained the chronology proposed in the original publications; still, the reader should bear in mind that many structures are not intrinsically dated and that there is a possible, even probable, margin of error that does not appear as such in our reasoning.

As the reader will have understood, I will focus on crop production, since evidence for specialization and intensification of animal husbandry is much more conclusive (see now Groot 2016). I will start by discussing the shape and size of granaries, trying to show that although larger storage facilities did appear in the second and third centuries CE, there was no straightforward evolution from small buildings in the Iron Age to big ones in the Roman period. I will then tackle the tricky problem of their storage capacity, which has been much overestimated in previous works. In my view, it is doubtful that the emergence of bigger structures in Roman times actually points to increased agrarian production. This will lead me to advance a new hypothesis concerning the changes observed in the Batavian countryside in the first three centuries CE.

1 Romanizing the Granaries?

Underground silos are not found in the Rhine area, for obvious practical reasons: the water table is high and flooding frequent, making this storage method highly unsafe. For mid- to long-term storage, above-ground granaries built on wooden posts have always been the preferred method. As said previously, there is clear continuity between the Iron Age and the Roman period. Kees Schinkel (1998, 255) designed a typology for Oss "Ussen", that was later adapted by Vos (2002, 49–50) for Wijk bij Duurstede "De Horden". It consists of three groups and is now widely used in the Netherlands (Table 6.1).

Group 1 consists of granaries on two rows of posts (generally 4 or 6), group 2 on three rows of posts (generally 9 or 12). These first two groups are known at least from the beginning of the Iron Age in the eighth century BCE and are

TABLE 6.1 Typology of granaries on post used in the Netherlands
BASED ON SCHINKEL 1998 AND VOS 2002

Sources	Schinkel 1998	Vos 2002
Type 1: 2 rows of posts		
4 posts	IA	1a
6 posts, narrow	IB	1b
6 posts, large	IC	1b
7+ posts	ID	1c
Varia	IE	1d
Type 2: 3 rows of posts		
9 posts	IIA	2a
12+ posts	IIB	2b
15+ posts	–	2c
Mid-row posts only on short sides	IIC	2d
Varia	–	2e
Type 3: *horrea*		
Floor on 9 posts surrounded by wall	IIIA	(3a)
Floor on 12 posts surrounded by wall	IIIB	(3a)
Posts in trench	(III)	3a
Varia	–	3b

found during the Roman period and beyond. Group 3 is more diverse and comprises storage structures generally referred to as *horrea* in the Dutch literature, i.e. large granaries with plans not known in the Iron Age, and therefore perceived and dated, most of the time with good reasons, as Roman.

The commonly accepted idea is that, during the Principate, granaries got larger, and at the same time boasted a Romanized architecture, apparently inspired by military storage buildings (Groot et al. 2009, 239 fig. 5, illustrates this idea quite clearly). This is indeed the general tendency; if one compares for instance the floor area of granaries from Oss "Ussen" from the Early Iron Age (8th-5th c. BCE) to the Roman period (1st–3rd c. CE), it is very clear that granaries of 20 m^2 or more only appear in the Roman period (Fig. 6.2). Evidence from other sites allows for more precision: granaries of 20 m^2 appeared around the mid-first century CE; granaries of 50m^2 appeared at the end of the century only, and the biggest granaries are generally dated from the mid-second

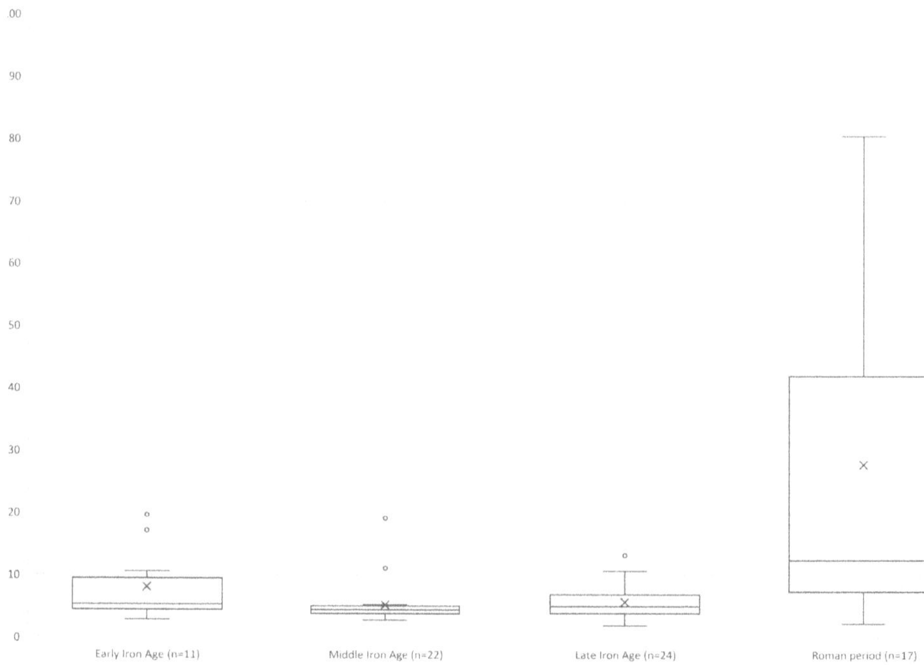

FIGURE 6.2 Boxplot of the surface in square meters of excavated granaries in Oss "Ussen". Quartile calculation includes the median. The cross indicates the mean.
S. MARTIN, BASED ON DATA FROM SCHINKEL 1998

century onwards (for a general overview in northern Gaul, see Habermehl 2013, 148–52).

However, a closer look at the data shows there is no linear progression from small four-post granaries to big Roman *horrea*. As the authors of the paper in the *JRA* themselves noted, large storage buildings are not to be found on every site (Groot et al. 2009, 249). For instance, in Geldermalsen "Hondersgemet", the biggest granaries from the first three centuries CE were never bigger than ca. 32.5 m^2 and most of them were in the 10 to 20 m^2 range characteristic of the Iron Age. Small granaries never disappeared and are found throughout the Roman period, sometimes alongside larger facilities (see e.g. for Tiel, Heeren 2009, 179–81, figs 93–95). Since they are also harder to date, their number for the Roman period is likely to be underestimated on several sites. Finally, on several sites, there is actually little to no evolution in terms of granary size and architecture, from the Iron Age to the Roman period (see e.g. Cuijk "De Nielt" or, for the first two centuries CE, Houten "Molenzoom" [cat. 307]).

In addition, the classification of most of the bigger granaries within a single "*horrea*" category is problematic. In their respective typologies, Schinkel and Vos adopt differing criteria to define them. For Schinkel (1998, 255), a granary is a *horreum* when a wall surrounds the floor (see fig. 6.3, granary B11 from Tiel and fig. 6.4, granary S309 from Oss). Vos (2002, 50) has two sub-types: the first comprises granaries with posts embedded in trenches, a technique generally considered as Roman, found for instance in military camps (see again fig. 6.3 and 6.4); the second is merely defined as "*diversen*", which is not very satisfactory. But it has the merit of making clear that, when there is no clear Roman architectural trait, it is size that acts as a criterion of Roman-ness. The case of granary 47 in Tiel "Passewaaij" is a good illustration of this (fig. 6.3): in the Dutch publication, it is defined as a *horreum* (Heeren 2009, 176, 311), which is translated as "Roman style granary" in the *JRA* paper (Groot et al. 2009, 243). But granary 47 does not fit in Schinkel's group III, nor in Vos' sub-type 3A (with posts in trenches); it is of course possible to classify it as Vos 3B, but structurally it belongs to group 2, i.e. granaries on three rows of posts (see e.g. the undated granary S455 from Oss for a similar plan, with similar dimensions: Schinkel 1998, 258 fig. 279).

Granary 47 from Tiel "Passewaaij" clearly shows that architectural traits perceived as Roman and size do not always go hand in hand. We also know cases of granaries that would be classified as *horrea* according to Vos, because the supporting posts are entrenched, but which are actually very small. This is the case for instance in Zaltbommel, where five granaries correspond to Vos 3A *horrea*: the largest has an area of 30.25 m^2, three measure 16 m^2 (thus within the range attested during the Iron Age) and the smallest… less than 3 m^2! Interestingly, the smallest of the five has been dated 100–175 CE whereas the others were out of use by 150 CE (Veldman and Blom 2010, granaries 4, 7, 12, 13, and 22; cat. 314).

What is more worrying is that granaries with entrenched posts are also known north of the Rhine, sometimes very far from the Roman frontier, for instance in Archsum, near the German border with Denmark. The structures are dated to the second century CE (Kossack, Harck, and Reichstein 1974; Kossack and Baessler 2001; cat. 300). Although Georg Kossack considers that the buildings were modelled after Roman prototypes, whether the presence of foundation trenches for posts is a marker of Romanization is doubtful, all the more so when one considers that this technique is also found in Iron Age Britain and Inner Gaul, although in this case with only two rows of posts (Guillier et al. 2015, 222, 223 fig. 37, 224 tab. 17, with earlier references).

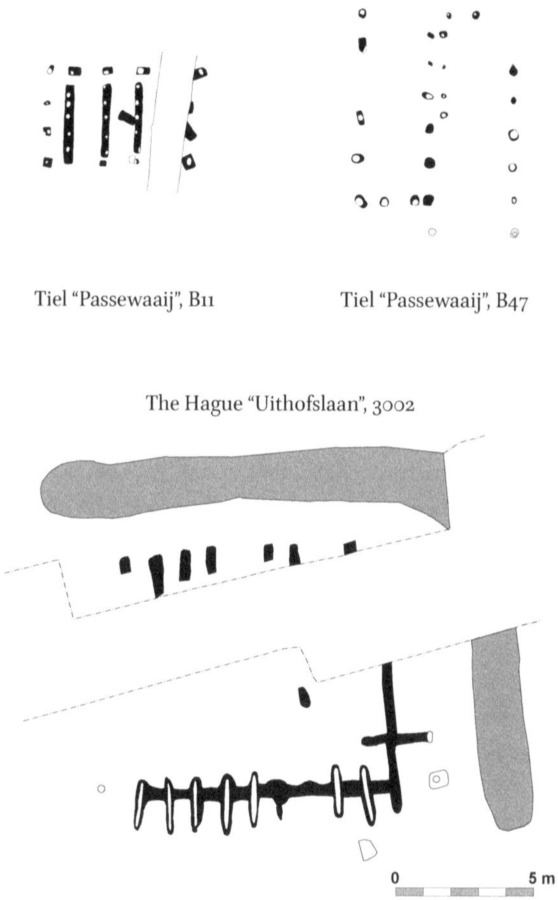

FIGURE 6.3
Granaries classified as "*horrea*" in the typologies from Schinkel and Vos. All plans at the same scale
FOR TIEL, HEEREN 2009, 179 FIG. 93 AND 180 FIG. 94; FOR THE HAGUE, PAVLOVIC 2011, 76 FIG. 3.16, WITH MODIFICATIONS

2 Calculating Storage Capacities

To sum up, without denying that some transformations in shape and size are obviously to be found in granaries from the Roman period, it must be stressed that there is neither a linear evolution from small to large storage facilities, nor towards a more Roman architecture. To go further, it is necessary to take into account all granaries and to investigate the storage capacities of the buildings in more detail. As written earlier in this book, however difficult it is, we have no other way of calculating the extent of potential surpluses.

The calculations in the *JRA* article are simple: for a given occupation period, the area of all structures identified as granaries is multiplied by the density of stored cereals (hulled grain is assumed, which is consistent with archaeological finds).[2] Of course, the authors are well aware that there is no straightforward relation between the floor area of a granary and how much could be stored in it (see Part I in this volume). They insist mainly on the lifespan of wooden granaries which, at an estimated 15 to 25 years, is less than the estimated lifespan of houses.

However, the proposed calculations are problematic. First of all, the density proposed for grain, 280 kg/m^3, is at odds with every other estimate found in the literature. It appears to stem from measurements made by Kooistra during her doctoral research, which gave a density of 367 kg/m^3 for spelt in chaff and 272 kg/m^3 for emmer in chaff, both after a single threshing (Kooistra 1996, 98 and 318, and not 66–8 as indicated in Groot et al. 2009, 235 note 22). These are very low values, in particular for emmer: P. Ouzoulias (2006, 174) gives a mean density of 460 kg/m^3 for hulled emmer. True, this last figure is quite high compared to the ranges found in other publications (Saunders 1904: 375–474 kg/m^3; Stallknecht, Gilbertson, and Ranney 1996: 360–440 kg/m^3); but even the lowest figures are much higher than Kooistra's measurements, which were taken after the first threshing. Furthermore, this is considerably lower than the density of barley, which probably constituted a large part of the stored crop (Plin., *Nat.* 18.62 gives a density of 560 kg/m^3 for hulled barley, consistent with modern data: Ouzoulias 2006, 176; Charrondiere, Haytowitz, and Stadlmayr 2015, 9).

Second, the difference between the total surface area and the usable surface area is not taken into account, notwithstanding the fact that one half of the largest storage facility in their corpus, granary 47 from Tiel "Passewaaij", seems to have been empty when the building burnt (Kooistra and Heeren 2007: grain was only found in the postholes of the southern half of the structure). This probably accounts for the excess of storage space they reconstruct. Indeed, they compare the results of their calculations with the reconstructed grain requirements of the inhabitants. The reconstructed quantity of stored grain is generally much greater than the reconstructed quantity of required grain (Groot et al. 2009, 242 table 1). So much so, in fact, that the granary cannot have been used for crop storage only, as the authors themselves note. They therefore propose that granaries were also used to store other products, such as animal fodder (Groot et al. 2009, 249. For the storage of fodder, see also Groot and Kooistra 2009; Vossen and Groot 2009). However, the idea that fodder accounted for some of the extra storage space is not supported by the article's own data. Indeed, the main use of fodder is to feed stabulated livestock. If the

2 For the western Lower Rhine, there has been one attempt by Otto Brinkkemper at calculating storage capacities before Groot et al. 2009, limited to two granaries belonging to Schinkel's group III (Brinkkemper 1993, 149–50; Brinkkemper et al. 1995, 163). Although it is cited in subsequent work, as a whole the method has not been applied to other sites and I will therefore not discuss it here.

growth of storage space corresponded to growing quantities of fodder, we would thus expect a similar growth in the number of stalls. However, stabulation space remains constant through the centuries, making extra storage of fodder unlikely (Groot et al. 2009, 242 table 1; calculation of stabulation space is explained at 235).

Since no significant change has been detected in the archaeobotanical records, whether in crop types or in the ratios between the various cultivated plants, a solution to the problem is not likely to come from this direction.

One way out is to revise the calculations and try to rectify some of the issues identified above, using the model proposed in Chapter 3 of this book. I will use hulled barley as the main crop, with a density of 560 kg/m³. This is denser than hulled wheat, and barley chaff appears to be lighter than that of emmer: in this respect, the following calculations should be taken as rather high estimates (Brinkkemper 1993, 149, for the proportion of chaff to grain: after a first threshing, chaff is estimated at only 1% of the total weight [note that Brinkkemper uses much higher densities]. For spikelets before threshing, we should probably assume 10 to 20% of the total weight). Usable surface area will be equated with 70% of the total area, with a storage height of 30 cm; this last figure is rather low, but is attested in Amiens "ZAC Cathédrale" (cat. 250) for hulled spelt (see Chapter 3). No attempt is made here to quantify post-harvest losses. The sample of sites has been expanded to include data not discussed in Groot et al. 2009. Furthermore, I have used occupation periods as defined in the original publications, rather than half-centuries as in the paper; this makes it more difficult to study Wijk bij Duurstede "De Horden", since most granaries are dated to two phases or not dated at all (this is the case, for instance, of the larger granaries).

The newly reconstructed storage space is compared with the original grain requirements proposed in the article (Groot et al. 2009, 234–35: a family of 6.5 persons per house, needing 1178.6 kg of grain each year to meet 70% of its protein requirements; the reserve is assumed to be 50% of the grain needed for consumption, and sowing-seed, 20% of consumption + reserve). Delving more precisely into these matters may have been useful, but seemed unnecessary here. It must be kept in mind that we are only dealing with estimates; although it does not appear in the figures, the reader should remember the margin of error might be substantial. As a benchmark against which to evaluate the relative value of various storage capacity calculations, the estimated grain requirements as published in 2009 are therefore good enough. Furthermore, it will be easier to show that the storage capacities were overestimated in 2009 by using the same grain requirements as in the original paper (Table 6.2).

Although the sample remains small, some tendencies are nevertheless clear. In Geldermalsen "Hondersgemet" and Tiel "Passewaaij", the relation between

TABLE 6.2 Ratio between needed and available storage capacity on various sites from the western Lower Rhine region (for calculation method and sources, see text).

Site	Phase	TPQ	TAQ	N hous
Cuijk "De Nielt"	F	200	270	4
	J	80	130	5
Geldermalsen "Hondersgemet"	1	200 BCE	50 BCE	3
	2	50 BCE	50	5
	3	50	120	5
	4	120	270	5
	5	270	350	1
Groesbeek "Hüsenhoff"	1	175	225	1
	2	225	250 (270 ?)	1
Houten "Molenzoom"	1	20	50	1
	5	140	170	1
	6	170	200	1
Houten "Terrein 8A"	1	1	40/50	2
	2	40/50	80/90	2
	3 to 5	80/90	200	2
Rijswijk "De Bult"	Ic	90	120	2
	Id	120	150	3,5
	II	150	210	3,5
	IIIa	210	240	3,5
The Hague "Uithofslaan"	1 and 2	60	120	4
	3	120	150	3
	4 and 5*	150	210	2
Tiel "Passewaaij"	2	50 BCE	50	4
	3	40	150	9
	4**	150	220	5
	5	210	240	2
	6	240	270	1
	7	270	350	2
Wijk bij Duurstede "De Horden"	1	1	50	8
	2	50	70	8
Zaltbommel "De Wildemann"	A1	1	50	2
	B	75	150	2,00

* no house excavated for phase 5

** 2 granaries have incomplete ground plans

Nb inhabitants	Storage capacity needed in m² (calculations according to Groot et al. 2009)	Storage capacity needed in m² (calculations according to this chapter)	Storage space available in m²	Ratio storage capacity available/ needed
26	21,05	50,11	20,19	0,40
32,5	26,31	62,64	29,00	0,46
19,5	15,78	37,58	156,07	4,15
32,5	26,31	62,64	56,00	0,89
32,5	26,31	62,64	73,38	1,17
32,5	26,31	62,64	92,60	1,48
6,5	5,26	12,53	97,85	7,81
6,5	5,26	12,53	51,92	4,14
6,5	5,26	12,53	190,26	15,19
6,5	5,26	12,53	17,32	1,38
6,5	5,26	12,53	14,81	1,18
6,5	5,26	12,53	11,76	0,94
13	10,52	25,06	10,80	0,43
13	10,52	25,06	9,61	0,38
13	10,52	25,06	22,75	0,91
13	10,52	25,06	11,03	0,44
22,75	18,42	43,85	31,50	0,72
22,75	18,42	43,85	93,50	2,13
22,75	18,42	43,85	64,66	1,47
26	21,05	50,11	66,45	1,33
19,5	15,78	37,58	28,75	0,76
13	10,52	25,06	87,40	3,49
26	21,05	50,11	193,49	3,86
58,5	47,35	112,75	137,86	1,22
32,5	26,31	62,64	106,92	1,71
13	10,52	25,06	28,72	1,15
6,5	5,26	12,53	4,40	0,35
13	10,52	25,06	9,36	0,37
52	42,09	100,22	74,70	0,75
52	42,09	100,22	127,45	1,27
13	10,52	25,06	26,30	1,05
13,00	10,52	25,06	99,25	3,96

available and required storage space is of 4:1 (in favour of the former) during the first phase, which corresponds to the Late Iron Age on the first site, and to the period from 50 BCE to 50 CE on the second. This is probably due to the construction of numerous small granaries with a short lifespan, a situation that is also found in the second half of the first century CE in De Meern "Oudenrijnseweg" (Luksen-IJtsma 2009; cat. 303). On this site, 20 granaries are associated with a single house; however, 18 are concentrated in two locations, next to or within an enclosure interpreted as a paddock, with several granaries overlapping and at least two rebuilt on the same spot. Foodstuff was probably stored in the two granaries closest to the house, clearly separate from the other ones.

In Tiel "Passewaaij" and Geldermalsen "Hondersgemet", the first century CE sees an apparent decrease in available storage capacity, which now matches required storage capacity. This ca. 1:1 relation can be observed on other sites, making it likely that the 4:1 relation from the previous period was indeed an artefact. There is little change between the first and second centuries (Houten "Molenzoom"; Tiel "Passewaaij", periods 3 and 4; Geldermalsen "Hondersgemet", periods 2 to 4; no perceptible evolution in Wijk bij Duurstede "De Horden", although the proportion of undated granaries should not be forgotten, nor in Houten "Terrein 8A" [cat. 308]; in the coastal region, see Rijswijk "De Bult", period I, with a very low ratio [cat. 25], and The Hague "Uithofslaan", periods 1 and 2 [cat. 302]). Some sites do see an increase in the second century, but given the margin of error, it might not be significant. It is only in the last phases, at the end of the second or during the third century, that available space greatly exceeds required space (Groesbeek "Hüsenhoff", period 1 and mostly period 2 [cat. 305]; a similar evolution is seen in the coastal region in Rijswijk "De Bult", periods II and III, and The Hague "Uithofslaan", periods 4 and 5; in The Hague "Wateringsveld", the largest granary [structure no. 112] is dated to the latest period). This change is limited to a number of sites (in the same period, available space is stable then decreases markedly in Tiel "Passewaaij", periods 4 to 7; in Geldermalsen "Hondersgemet", the increase is only perceptible in period 5, which constitutes a rare case of post-270 CE occupation in the region).

On the whole, then, on most sites and during most of the Roman period, there appears not to be an excess of storage space. If this new reconstruction is to be accepted, the need to explain it, by storage of fodder or other products, disappears – but so does the existence of significant surpluses. However, this conclusion is actually more in line with other studies written before or contemporaneously with Groot et al. 2009. For instance, in 1996, Kooistra wrote in her dissertation, speaking of the Kromme Rijn area:

> With such a dense population, it is hardly possible to produce more than is required. From the organic material it appeared that more was probably produced than the population needed. With regard to the foodstuffs, this production must have been of marginal importance for the non-agrarian population in the Kromme Rijn area. (Kooistra 1996, 72)

Some years earlier, Guus Lange, writing about Wijk bij Duurstede "De Horden", was even clearer: "The remains from De Horden do not allow the conclusion that the emphasis on arable farming increased. The minor importance of cereal remains in [later periods], compared to [the first period of occupation], points rather to the contrary" (Lange 1990, 140). Another important argument is that the available arable land was limited to a small part of the landscape, namely the stream ridges, which area could not be expanded (this is well demonstrated for Tiel "Passewaaij" in Groot and Kooistra 2009). One could argue for a more intensive exploitation of stream ridges, but this means a larger workforce, and the size of the settlements does not indicate any population boom (not to mention that part of the new surpluses would have fed the extra population). The intensification of production seemingly brought by the Roman conquest is not to be looked for in crop production, as Groot (2016) has shown convincingly in her latest monograph, confirming a suggestion by Kooistra (1996, 71–73).

3 A "Romance of Storage" in a Non-Villa Landscape?

If all the conclusions from the 2009 article cannot be upheld in their original form, it remains true that a gradual trend towards larger storage buildings (not necessarily correlated with an increase in the overall storage capacity) can be discerned from the mid-first century CE onwards, culminating, on some sites only, with the erection, during the second half of the second century and in the third century, of granaries (sometimes greatly) exceeding the storage space needed by the inhabitants. If this is not related to a corresponding growth in grain or fodder production and storage, how is it to be explained?

Although it did not have their favour, for lack of existing research, another hypothesis was put forward by the authors: sites with large granaries could have acted as collecting sites for a given area (Groot et al. 2009, 249). This claim is not supported so far by archaeological data, since the construction of large granaries on some sites does not correspond to a reduction in storage space at other settlements. But the underlying idea, namely that the control over agrarian products could vary between various settlements, and it may be added, between different houses of the same settlement, is appealing and can be substantiated by archaeological evidence.

The evolution in the shape and size of granaries must be put in perspective with contemporary changes on the western Lower Rhine. The emergence of large granaries follows the same chronological trend as the structuration of settlements with ditches. Although settlement nucleation started in the Late Iron Age, ditches were still uncommon during this period and enclosed settlements almost inexistent (Gerritsen 2003, 181–89; Arnoldussen and Jansen 2010; Mathiot 2012, who shows that the lack of enclosures does not mean settlements were not organized). Settlements enclosed by ditches become commoner around the mid-first century CE, when the first granaries with a floor area greater than 20 m^2 are attested. The largest storage buildings appear in the second and third centuries, when enclosed settlements are the norm; at this period, ditches also divide the settlements internally, and they sometimes extend into the neighbouring areas, marking fields and other boundaries (Heeren 2009, 241–50; Vos 2009, 104–16, focused on Wijk bij Duurstede; on Oss in the Roman period: Wesselingh 2000, Jansen and As 2012; for the coastal region, Londen 2006, 183–221). Whether the ditch systems are the result of direct Roman intervention or not will not be discussed here, for it would take us too far. But scholars agree that ditches and enclosures, in addition to practical uses for irrigation or drainage, reflect changes in land management and perhaps ownership, and also have a symbolic meaning (particularly for ditches inside the settlements).

Indeed the development of ditches generally goes hand in hand with growing signs of social distinction inside the settlements, with one (sometimes more) house(s) clearly assuming more importance. In the Batavian area, they often take the form of the so-called "porticus-houses", i.e. traditional houses with a portico on one or more sides. The interpretation of these porticus-houses is much debated: according to Vos, the portico is an influence from the military sphere, whereas Harry van Enckevort sees them as a local type of villa (Vos 2009, 237–51, summarized in 2015, 453–54; Enckevort 2012, 281–87; see also Groot 2011 for links between porticus-houses and specialization in horse breeding). But everyone agrees that they were special houses, very likely the seat of the leading individual or family of the settlement (see Jansen and Fokkens 2010 for the porticus-houses as *Herrenhöfe*). Porticus-houses are not attested in the coastal area but similar tendencies are to be found; in Rijswijk "De Bult", for instance, the principal house was rebuilt in stone on the same spot during the second century, a clear sign of stability and continuity (Bloemers 1978: house 17 was built in period Id, replaced by house 18 in period IIa; during period IIb,

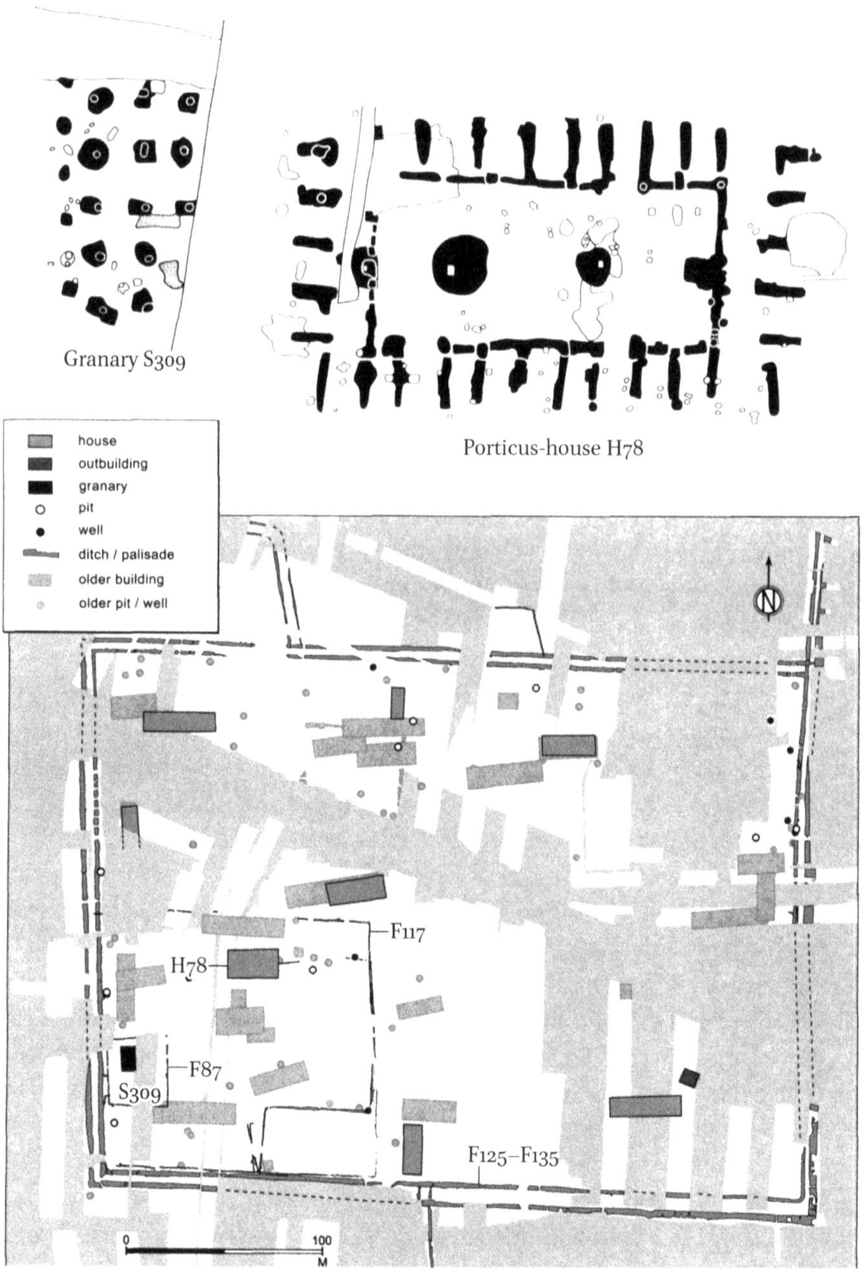

FIGURE 6.4 Oss "Westerveld", phase 4 (70–125 CE). Association between porticus-house H78 and large granary S309
WESSELINGH 2000, 83 FIG. 82A, 113 FIG. 113C, AND 165 FIG. 190, WITH MODIFICATIONS

house 19 was built with a stone cellar; during the third century, i.e. periods IIIa/IIIb, house 19 was rebuilt with stone foundations).

There are a few cases where the link between special houses and large granaries is well attested. One of the best examples probably comes from phase 4 of the settlement of Oss "Westerveld", dated between 70 and 125 CE (Fig. 6.4). Inside the big square ditch that encloses the whole settlement (F125–F135), is another smaller enclosure (F117) around a porticus-house (H78). Inside this smaller enclosure is yet another one (F87), with a granary inside (S309). In addition to its architecture, the porticus-house H78 is singled out as special by a variety of finds: two keys, two of the three coins found on the whole settlement, most of the iron and bronze objects of the settlement, a fair share of the glass vessels, and a large concentration of tiles (Wesselingh 2000, 71–169 and 214–21). The granary is obviously too large for just one house: at 80.5 m^2, it is the largest of the 479 storage buildings, although house H78 is far from being the largest (Schinkel 1998, 250–51 for a catalogue of houses, 262–66 for granaries; granary S309 is a *horreum* belonging to his type IIIb). Do we have to imagine that its use was effectively restricted to this one house, even if it was not used to its full capacity? Or did the dwellers in house H78 control and store produce harvested by the other inhabitants of the settlement? To a certain extent, both hypotheses are probably true.

Contemporary with Oss "Westerveld", two settlements from Houten "Doornkade" [cat. 306] and Houten "Tiellandt" offer the same association between *horreum* and porticus-house. In "Tiellandt", the area is notable for the discovery of stamped tiles and of an inscribed stone altar, one of the only two stone inscriptions found in the region of Houten (Vos 2009, 118–27 for Doornkade, 133–41 for Tiellandt). In Risjwijk "De Bult", in the second half of the second century, a large *horreum* (no. 26) is close to the principal house (no. 19); although it is not in the same enclosure, it is partially fenced off from the closest house (Bloemers 1978: during the first half of the third century, granary 26 is replaced by granary 27. Both storage buildings are close to structures 5 and 6, interpreted by Bloemers as temples. But some scholars interpret building 5 as a storage facility rather than a temple; see Habermehl 2013 with previous literature. For some reason, the identification of building 6 is not discussed). The same situation occurs nearby in The Hague "Uithofslaan" during period 4 (150–180 CE) where two buildings interpreted as storage facilities (*horrea* 3002 and 3003) are enclosed by a ditch inside a wider enclosure, with house 1003 lying inbetween (Pavlovic 2011, 371 fig. 11.6 for an overview of the settlement during period 4. It should be noted that the plan of building 3002 is quite puzzling. The only comparison known to the author of the report comes from a very limited excavation and does not allow for a positive identification of the building's function: Heeringen, Lauwerier, and Velde 1998, 23–25).

This development is not restricted to the western Lower Rhine region. In the neighbouring *civitas* of the Menapians (corresponding roughly to modern-day West-Flanders), Wim De Clercq has shown that granaries are part of what he calls "a socially constructed spatial layout displaying a successful harvest and hence the prosperity of the family" (De Clercq 2011, 242), which is particularly clear on the site of Bruges "Refuge" (cat. 301). This settlement is organized along an axial perspective, not dissimilar to the one found on *villae*, and granaries are clustered around the main house located at the end of the courtyard (De Clercq 2009, 225–28, 2011, 240–45. For axial *villae*, see Ferdière et al. 2010; Roymans and Habermehl 2011). This is all the more interesting since soils are poorer than in the Meuse-Rhine delta and possibilities of surpluses even lower. Moreover, the same development can be studied in the Gallic "villa landscape": for Northern Gaul and the Rhine area, this has been made very clear by the comparative and diachronic study of Diederick Habermehl (2013, 153–55).

On the basis of the archaeological evidence presented above, one gets the impression that the increase in size of granaries is the sign of increased power, that could have rested on an increased control over crop production and storage in the settlement. This calls to mind remarks made by Nicholas Purcell about Italian *villae*. Purcell famously wrote of "*villae* and the romance of storage", underlining how visibly storage facilities (granaries but also cisterns) were displayed, and Astrid van Oyen has recently explored the theme further, speaking of "conspicuous production" (Purcell 1995; Oyen 2015). Although the latter may underestimate the impact of technological and economic factors in designing a storage building, they both stress rightly that this was in part a cultural choice and not only an economic necessity.

Of course, the situation of the Italian countryside cannot be transposed as such to the Low Countries. But recent studies may have put too much emphasis on the specificity of the Batavian area, and more generally of the non-villa landscapes. For instance, H. van Enckevort has argued in his dissertation that settlements with porticus-houses and other non-indigenous features should actually be interpreted in the light of contemporary *villae*, of which they could be a local variety (Enckevort 2012, 281–90). Indeed, one could argue that some of the changes observed in rural settlements are linked to the process of municipalization that shaped the Batavian *civitas* from the Augustan period onwards. By municipalization, we mean the process by which communities were integrated into the Roman world, by being given the status of *civitas* and the institutions that went with it and that tended, with time, to closely resemble Roman institutions (see Dondin-Payre and Raepsaet-Charlier 1999 for a thorough study of the topic). It must be stressed that, even if Tacitus (*Germ.* 29.2) speaks of the *antiqua societas* which united them with Rome and dispensed them from paying normal taxes in the first century CE, Batavians were municipalized exactly like the rest of the provinces of Gaul and Germany. They were probably organized as a *civitas* in the last years of the first century BCE, when Augustus created the province of *Germania* at the beginning or after the successful campaigns of Drusus (see Faoro 2015 with previous literature; the idea was already put forward explicitly in Wightman 1985, 55). The *ara Ubiorum*, was built at the same time in Cologne for the provincial cult and it is most likely that the Batavians sent delegates. In this first phase of municipalization, the *civitas* retained at least partly indigenous institutions, as demonstrated by the inscription from Ruimel (CIL XIII, 8771), dated to the first half of the first century CE: it mentions a *summus magistratus civitatis Batavorum*, who did not have Roman citizenship and bore a Germanic name, *Flaus Vihirmatis fil*(*ius*). It was only with the creation of the new town of *Ulpia Noviomagus* around 100 and the granting of municipal rights probably not long afterwards, that Batavians got truly Roman institutions, with a *curia* of *decuriones*, from which magistrates were chosen each year (epigraphic data are compiled in Raepsaet-Charlier 1999, 325–26; see Willems and Enckevort 2009 for an archaeological and historical synthesis). This gradual process of institutional Romanization certainly had an effect on Batavian society. Not everybody could be part of the ruling body of the *civitas* and, particularly after it became a *municipium*, magistrates were chosen from a small pool of families. This reinforced the social hierarchy and it is very possible that the process was reflected in the settlements in the countryside, with enclosures, porticus-houses and larger granaries appearing in some settlements only, while others remained very much as in the Iron Age (Enckevort 2012, 279–81, suggests that the "proper" *villae*, presented in chap. 11 of his study, were inhabited by members of the *ordo decurionum*). This would also explain why porticus-houses and big granaries disappeared after the Roman period, while "traditional" granaries and houses remained more or less the same down to the Middle Ages (for house-plans through history in the Netherlands, see Lange et al. 2014. As for granaries, it is more difficult to have an idea of their aspect based solely on the number of supporting posts; it is clear, however, that 4-post granaries remained the commonest type well after the Roman period: see e.g. Habermehl and Renswoude 2017 for Cuijk "De Nielt", with clear continuity from the Bronze Age to the Middle Ages). Porticus-houses and so-called *horrea* were typical of a particular socio-economic situation and vanished with it.

What, then, was the economic significance of large granaries in the Batavian area? If the arguments presented above hold water, these granaries should not be taken as indicators of surplus grain production. The Roman conquest has no discernable impact on the kind and the proportions of crop species produced in the area, and the shortage of arable land in the western Lower Rhine region makes any significant increase in production unlikely. The integration into the Roman empire did have an effect on agrarian production, but it was limited to animal husbandry. In this domain, archaeological evidence for specialization and surplus production is much more solid and convincing. Furthermore, animal husbandry is much better suited to the regional landscape. Additionally, it requires less labour than crop production, which is again consistent with the reconstructed demographics of the region.

On another level, the study carried out in this chapter has also proved interesting in highlighting some of the assumptions underpinning previous research on the subject. Once again, the link between Romans, the market economy and economic growth is patent. In the article commented on at length in this chapter, the only way considered for disposing of agrarian surplus is the market. Neither taxes in kind nor special levies are evoked. Furthermore, the idea that surpluses were produced during the Roman period has such a hold that the problem of land availability, well discussed in Laura Kooistra's previous research, is not raised.

Besides, it is clear that the Batavian *civitas* was not as special as some works make it appear. The fact that on the one hand, it is rather extensively mentioned by Tacitus, in comparison with other *civitates*, and that on the other hand, it is mainly studied by Dutch scholars, whose work is largely ignored abroad, even when published in another language than Dutch, has led to its uniqueness being overestimated. This is not to deny that it did have some specific traits; but when compared with other regions from the Three Gauls and the Germanies, it becomes clear that Batavians experienced the same overall change in municipalization, town planning or architectural development. In the countryside, changes became visible in the second half of the first century CE, with the first porticus-houses, and a few rare *villae* around Nijmegen, at the end of the century, followed by a growing demarcation of land by ditches and increased monumentalization (although very limited by Roman standards) in the second and third centuries: all in all, this is not very different from what happened elsewhere.

This last remark leads to a final question: if the interpretation of granaries in a non-villa landscape put forward in this chapter has some truth to it, could it not be applied to the villa landscapes? Should not the monumentalization of granaries in Inner Gaul or in Upper Germany be read along the same lines? And if so, how should we interpret the relation between their increased capacity and a supposed increase in productivity?

Des greniers ruraux aux greniers militaires et urbains. Les enjeux historiques d'une enquête archéologique

Michel Reddé

Comme l'a souligné à juste titre Stéphane Martin dans l'introduction de cet ouvrage, le regain d'intérêt actuel pour les structures de stockage de l'Antiquité n'est pas seulement une lubie d'archéologues en mal de classement typologique[1]. Il ne s'agit rien moins, en effet, que de tenter d'aborder des questions aussi complexes que la capacité productive des campagnes à un moment donné du temps, mais aussi celle des circuits de distribution. J'englobe dans ce terme à la fois le marché libre mais aussi les réquisitions, la fiscalité et la question des réseaux, privés, publics, commissionnés qui gèrent les relations entre producteurs et consommateurs. Or ces derniers ne sont pas tous égaux : l'armée, les centres urbains, grands ou petits n'ont pas nécessairement les mêmes besoins de ravitaillement et n'obéissent pas forcément aux mêmes logiques économiques.

Jusqu'à une époque récente, la question du stockage a surtout intéressé les historiens de l'époque romaine, notamment les spécialistes du monde méditerranéen, préoccupés de comprendre comment fonctionnait l'annone de la mégapole romaine, un problème qui a focalisé l'attention et détourné le regard des régions périphériques de l'Empire, particulièrement celles de l'Europe tempérée[2]. Les difficultés croissent en effet à mesure qu'on s'éloigne des secteurs pour lesquels on ne dispose plus de sources textuelles ou épigraphiques et où toute l'interprétation repose sur l'archéologie, car même les enquêtes de terrain les plus sophistiquées ne sont pas toujours capables de nous apporter des informations très précises sur la nature des denrées produites dans les campagnes et conservées dans des entrepôts souvent arasés jusqu'aux fondations. On peut essayer de rappeler brièvement quelques-uns des problèmes qui se posent à nous.

1 Je ne reprends pas ici l'historique de la question ni l'exposé des problématiques, tels qu'ils ont été énoncés dans ce texte introductif.

2 C'est ce que l'on constate toujours quand on consulte la base en ligne de l'ANR Entrepôts (<http://www.entrepots-anr.efa.gr>, consulté le 7 juin 2018).

a. En Gaule, notre information sur les structures de stockage est nettement meilleure dans les zones de production que dans les zones de consommation urbaine. Seuls font exception les greniers militaires, relativement mieux connus, qui nous permettent d'évaluer les besoins propres de l'armée. Mais que savons-nous, par exemple, des entrepôts d'une capitale provinciale comme Cologne, *a fortiori* de chefs-lieux de cités comme Saint-Quentin, Metz, Langres … ? Et ne parlons pas des agglomérations secondaires[3]. À l'inverse, en Italie, et notamment à Rome, à Ostie ou dans le golfe de Naples, notre connaissance des capacités de stockage destinées à la consommation urbaine est bien supérieure, alors que celle des établissements ruraux n'est pas la mieux étudiée.

b. Comment apprécier la capacité productive des campagnes de la Gaule indépendante ? Lorsque, sur un établissement de l'âge du Fer, nous mettons au jour un petit nombre de greniers sur poteaux, avons-nous affaire à la totalité de la récolte ou seulement à la part réservée pour la consommation domestique et les semences de l'année suivante ? La question peut en effet se poser depuis que nous connaissons l'existence, pour l'instant encore très exceptionnelle, de sites concentrant de nombreuses petites unités d'entrepôts, comme c'est le cas à Entrammes, dans la Mayenne, mais dont la totalité des structures est loin d'être négligeable puisqu'on atteint une superficie globale d'environ 720 m^2 (Guillier et al. 2015; voir dans le présent volume la fig. 5.11 et les p. 102–104). On peut en effet s'interroger sur le rôle les agglomérations ouvertes de la Tène C et des *oppida* de la Tène finale dans la centralisation des denrées alimentaires produites dans les campagnes environnantes. On doit aussi évoquer le rôle des fermes dites "aristocratiques" de l'Ouest de la Gaule, où la capacité de stockage semble significative mais n'est pas nécessairement le signe assuré d'une possession individuelle : une propriété collective n'y est en effet pas exclue. La plupart du temps, nous ignorons donc sur quelle base foncière et sur quel type d'exploitation repose la quantité de produits conservés dans un endroit donné (voir *supra* Chapitre 4).

c. Une question exactement inverse se pose lorsque, à l'époque romaine, nous rencontrons l'existence de greniers à très grande capacité au sein d'un site rural, comme c'est le cas à Bad Rappenau "Maueräcker" (cat. 251) ou Bad Rappenau "Babstadt", près de Heilbronn (Baden-Württemberg, DE), ou encore à Biberist (CH ; cat. 85). Dans le premier cas, les auteurs, partant de l'hypothèse d'un domaine de 100 ha autour de la *villa*[4], considèrent que ce grand *horreum*

3 Dans son dernier ouvrage, Javier Salido Domínguez en a dressé, pour les Gaules et les Germanies, une liste courte mais le mélange des types de structures, d'agglomérations, de dates permet justement de constater combien la question est mal connue (Salido Domínguez 2017, 124–32).

4 Elle n'est fondée que sur l'absence d'autres vestiges connus autour de la *villa*.

de 608 m², sans doute à deux étages, si on prend en compte la présence de ses contreforts, est trop vaste pour la production d'un seul domaine, et ils concluent à une fonction de stockage central pour plusieurs établissements ruraux (Hartmann et Meyer 2001 ; Wulfmeier et Hartmann 2009). Cette hypothèse a fait école : dans un article paru récemment, Matthieu Poux et ses collaborateurs proposent ainsi, à partir de l'analyse du vaste *granarium* de Panossas (Isère), de reconstituer un système d'approvisionnement annonaire à destination du *limes* de Germanie en s'appuyant sur l'existence d'un réseau d'*horrea* à grande capacité situés près des grandes voies du réseau d'Agrippa, depuis le nord de la Narbonnaise jusqu'à la mer du nord (Poux 2017 ; cat. 134). Nous aurons à revenir sur cette question fondamentale : sommes-nous, dans ces cas-là, face à des greniers publics installés au centre d'une zone rurale pour collecter localement les denrées destinées au ravitaillement militaire, ou devant des *latifundia* privés, ou encore des domaines impériaux, pourvus de gigantesques *horrea* ?

d. La dernière question, qui a été longuement posée tout au long de cet ouvrage, est celle du volume de stockage de tel ou tel type de bâtiment, en fonction de son architecture mais aussi de son mode d'entrepôt des grains. Notre évaluation de la capacité productive des campagnes, de la superficie de l'*ager* cultivé dépend largement de la manière dont nous calculons cette contenance, un exercice dans lequel, il faut bien l'avouer, entrent une bonne part de spéculations et d'a priori historiques. Pour ne pas revenir une nouvelle fois sur ces calculs souvent aléatoires, nous aborderons ici la question en considérant seulement l'emprise au sol des bâtiments, sachant que celle-ci peut être doublée par l'adjonction d'un étage, et sans tenir compte du mode de stockage, en vrac, en caissons, en tas ou en sacs. Nous prendrons essentiellement en compte la différence relative de superficie, à technique identique de dépôt des mêmes produits.

Pour ce faire, nous avons volontairement sélectionné ici différents greniers empruntés respectivement au monde militaire, urbain et rural, afin d'en comparer les écarts, mais en conservant dans un tableau les volumes estimés par les auteurs et qui peuvent considérablement varier selon le mode d'entrepôt supposé (Tabl. 7.1, p. 142–144). On ne prend en compte, pour simplifier, que l'hypothèse d'un stockage des grains, tout en sachant bien que d'autres denrées pouvaient être conservées dans ces bâtiments (les données chiffrées d'estimation volumétriques sont empruntées à Blöck 2011-2012 ; Demarez et al. 2010 ; Kehne 2008 ; Poux 2017 ; Schubert 2016 ; Schucany 2006 ; Willems 1988).

On constate que la contenance des mêmes greniers diffère très sensiblement d'un auteur à l'autre, selon le mode de stockage et/ou l'architecture que l'on reconstitue, ce qui explique que nous ne prenions ici en compte que la superficie. En ne considérant que ce critère et à mode de stockage identique,

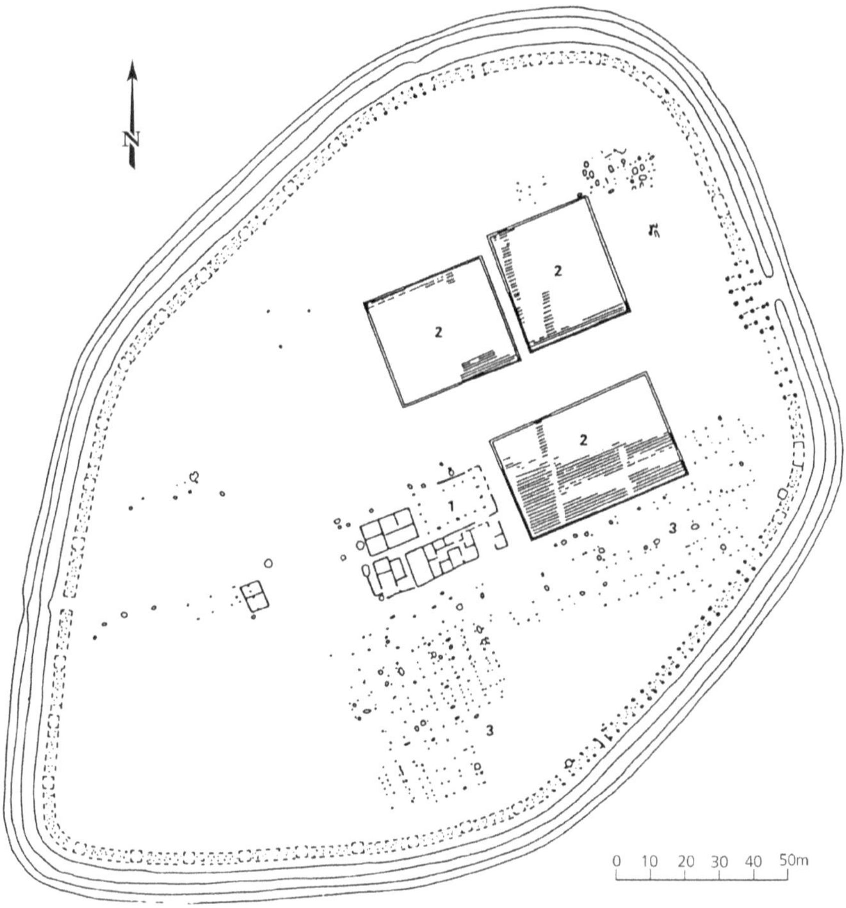

FIGURE 7.1 Le camp de Rödgen (Bad Nauheim, Hesse) et ses grands *horrea* (2)
REDDÉ ET AL. 2006, 217

les écarts entre les grands sites militaires de la conquête de la Germanie et les plus grandes *villae* connues sautent immédiatement aux yeux.

Les greniers de Rödgen, avec trois *horrea* occupant une superficie de 3457 m² (fig. 7.1), sont au moins 5,85 fois plus grands que celui de la *villa* de Biberist, 9,21 fois plus vastes que celui de Voerendaal (cat. 33). Le camp de Rödgen est généralement considéré comme lié aux opérations de Drusus, en 11-9 av. n. è. L'énorme "*Horreastadt*" d'Anreppen, sans doute mise en place pour la campagne de 9 de n. è., n'atteint pas cette capacité, avec "seulement" 2370 m², ce qui reste fort respectable, surtout si l'on y ajoute l'autre grand magasin à cour (fig. 7.2–3). En revanche un camp légionnaire classique comme celui de Neuss "Koenenlager", au IIe s. de n. è., possède des greniers d'une

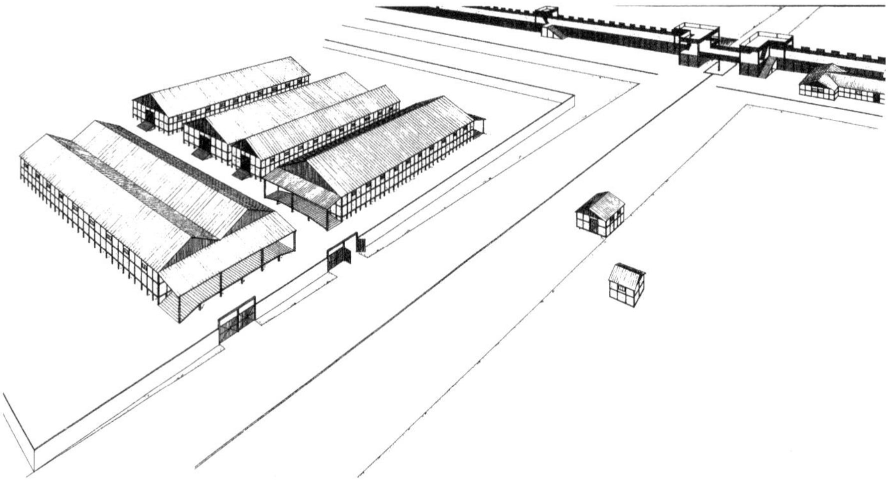

FIGURE 7.2 Delbrück/Anreppen : restitution de l'"*Horreastadt*" par J.-S. Kühlborn
REDDÉ ET AL. 2006, 111

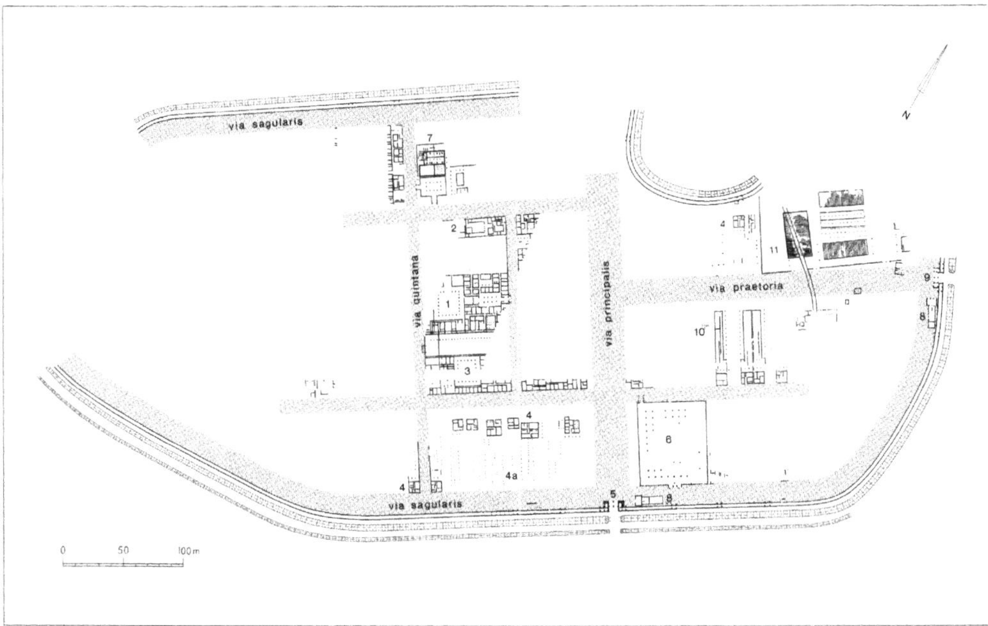

FIGURE 7.3 Delbrück/Anreppen : plan général, par J.-S. Kühlborn
REDDÉ ET AL. 2006, 261

superficie de 858 m² soit 1,5 fois la capacité de Bad Rappenau "Maueräcker" ou de Biberist.

Les camps auxiliaires disposent naturellement de capacités inférieures, mais parfois très hautes si on les compare à celles des *castra* légionnaires : ainsi est-on étonné de la taille des *horrea* de la Saalburg (489 m²) quand on la confronte à celle du "Koenenlager", dont l'effectif théorique était pourtant 10 à 12 fois supérieur. À Walheim II, dans la vallée du Neckar, le camp de *numerus* initial semble avoir été transformé en base logistique régionale, avec la construction de très vastes greniers, successivement G et F1, beaucoup trop grands pour satisfaire aux seuls besoins de la garnison restante (fig. 7.4). Bien clairement, même à une époque où les grandes expéditions de conquête avaient cessé, l'entrepôt des denrées nécessaires au ravitaillement militaire n'obéissait donc pas toujours à des lois de proportionnalité avec la taille de la garnison. On peut en outre supposer que le système avait passablement évolué au cours du temps : ainsi, dans le camp A d'Oedenburg, daté entre Claude et la fin du règne de Néron, l'*horreum* fouillé est fort limité (environ 48 m²) même s'il est accompagné, comme c'est souvent le cas, d'un magasin à cour beaucoup plus vaste, servant aussi au stockage de denrées autres que les grains (Reddé 2009, 120). La superficie probable du camp (sans doute autour de 2,5 ha) le place dans la catégorie supérieure des camps claudiens comparables, sans qu'on puisse affirmer qu'il s'agit d'un cantonnement auxiliaire car sa garnison semble mixte (Reddé 2009, 110–11). Les *horrea* d'Oberstimm, sur le Danube, sont nettement plus vastes (117 m²), alors que la superficie globale du *castellum* est plus faible (1,63 ha).

Ces différentes données doivent être confrontées avec celles d'établissements de la Tène finale dont la capacité totale de stockage est connue. À Jaux, en Picardie, par exemple, les trois greniers sur poteaux occupent une superficie d'environ 24/25 m², auxquels il faut ajouter la contenance d'un silo pouvant accueillir environ 2,8 t de blé, capable de nourrir 11 personnes pendant un an, soit la population probable de la ferme (Malrain et al. 1996). Si cet ensemble représente bien celui de toute la production annuelle, et en prenant pour hypothèse que le silo suffit à la consommation domestique (mais l'hypothèse inverse n'est pas exclue) cela implique qu'il aurait fallu environ 135 à 145 fermes identiques pour remplir les greniers de Rödgen, à hauteur et à mode de stockage identiques. Comme il n'était naturellement pas pensable de ne rien laisser aux producteurs pour se nourrir et ensemencer l'année suivante, on peut mesurer l'impact et l'extension régionale des prélèvements opérés sur les campagnes de Gaule intérieure pour fournir les *copiae* d'une partie de l'armée de Germanie, au début de la conquête. Dans le cas de Rödgen, il ne s'agit que d'une seule base logistique et nous ne savons pas si elle suffisait au ravitaillement de toute l'expédition.

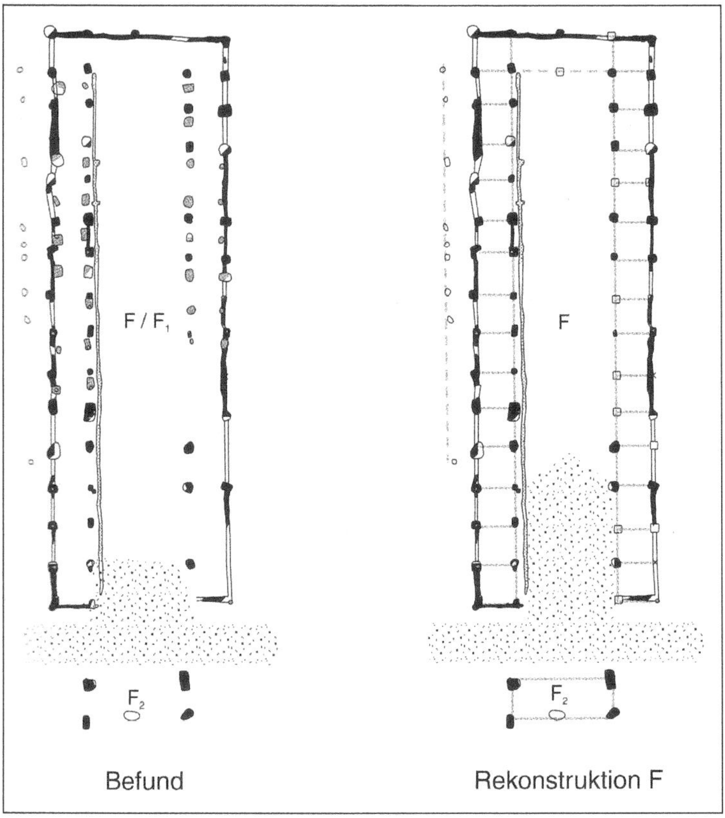

FIGURE 7.4 Le bâtiment F de Walheim, d'après K. Kortüm
REDDÉ 2011A, 127

Comparé à la superficie du grenier de la petite *villa* de Hambach 111 (cat. 97) dans les lœss fertiles des *Braunkohlenrewier*, soit 37,5 m², on ne constate pas de saut quantitatif important par rapport à Jaux qui comprend en outre un important silo. Les autres établissements de cette région, de taille généralement moyenne, disposent certes de capacités de stockage supérieures, mais celles-ci sont très inférieures à celles des très grands *horrea* qu'on a l'habitude de citer, et elles se rapprochent plus certainement de la réalité. En outre, on ne sait pas toujours bien dater ces greniers de manière précise et leur évolution spatiale dure parfois jusqu'au IIIe s. de n. è. (Schubert 2016).

Bien sûr ces évaluations et ces comparaisons sont probablement fausses dans l'absolu, d'autant qu'elles ne concernent pas des sites contemporains. Elles ne sont donc proposées ici que pour fournir un ordre de grandeur raisonnable, à partir d'une base de calcul homogène, et permettre de réfléchir à ce qu'implique, en termes économiques, la superficie d'une structure de stockage. L'archéologie observe par ailleurs une tendance continue à l'agrandissement des greniers ruraux, tout au long de l'Empire. Diderick Habermehl (2013) en a fourni une illustration suggestive à l'aide d'une double planche que nous reproduisons ici (fig. 7.5). Toutefois, cette planche peut conduire à des interprétations discutables si nous n'y prenons garde. L'accroissement de la superficie des greniers de Seclin, par exemple, signifie-t-elle une augmentation productive de l'ordre de 1 à 60 ? Évidemment non, sauf si l'on admet un accroissement considérable de la surface cultivée, donc une transformation de la structure foncière et probablement un phénomène de concentration. Dans le cas de Alle (cat. 125 et 221), mentionné plus haut dans le tabl. 1, il est peu probable qu'un même terroir ait pu, dans les conditions techniques relativement stables de l'Antiquité, passer d'une production stockable dans un grenier de 48 m^2, au début du Ier s. de n. è., à une capacité près de 15 fois supérieure, deux siècles plus tard. Cette modification considérable traduit évidemment autre chose qu'un accroissement des rendements et de productivité.

Les greniers urbains, en Gaule du nord, restent malheureusement mal connus. Les plus récemment fouillés et publiés, ceux de la rue Maucroix à Reims, ne sont pas dégagés dans leur totalité, de sorte que leur superficie totale est le fruit d'une estimation (fig. 7.6 et 7.7). On doit les confronter à ceux de Trèves, datables du début de l'Antiquité tardive, à un moment où cette ville devenait la capitale de la Gaule ; ils occupent une superficie de 2800 m^2 (Merten 2005). Il est donc douteux que ceux de Reims, au Ier s., aient pu atteindre une superficie au sol de près de 3250 m^2, comme le propose Maxence Poirier (2011; ici fig. 7.8). On serait là, en effet, dans une dimension proche de celle des *horrea* des grandes bases logistiques des campagnes de Germanie, destinées à approvisionner plusieurs dizaines de milliers d'hommes pendant des mois et supérieure à celle des *horrea* de Trèves.

Le cas des grands entrepôts mis au jour par Éric Binet à Amiens, toujours inédits, n'est pas moins douteux : les fouilleurs restituent une série de 10, voire 11 édifices, construits vers 80 de n. è., précédés par des portiques, et constituant des îlots (fig. 7.9). Mais il s'agit là d'une proposition hasardeuse, compte tenu de l'état de conservation réel des vestiges dont une grande partie est totalement arasée : la superficie restituée – 3000 m^2 – en ferait les plus grands de notre série et les situerait au niveau des grands entrepôts romains, une hypothèse sans doute excessive pour la petite ville qu'était Amiens au Ier s. : ainsi, à Rome, les *horrea Agrippiana* offrent-ils une dimension de 2750 m^2, les *horrea Lolliana* de 2500 m^2 (Bernardos Sanz et Virlouvet 2016, 82). L'identification comme greniers n'est d'ailleurs pas absolument certaine, bien que possible, mais ce n'est pas le lieu de discuter ici des autres hypothèses possibles. À l'autre

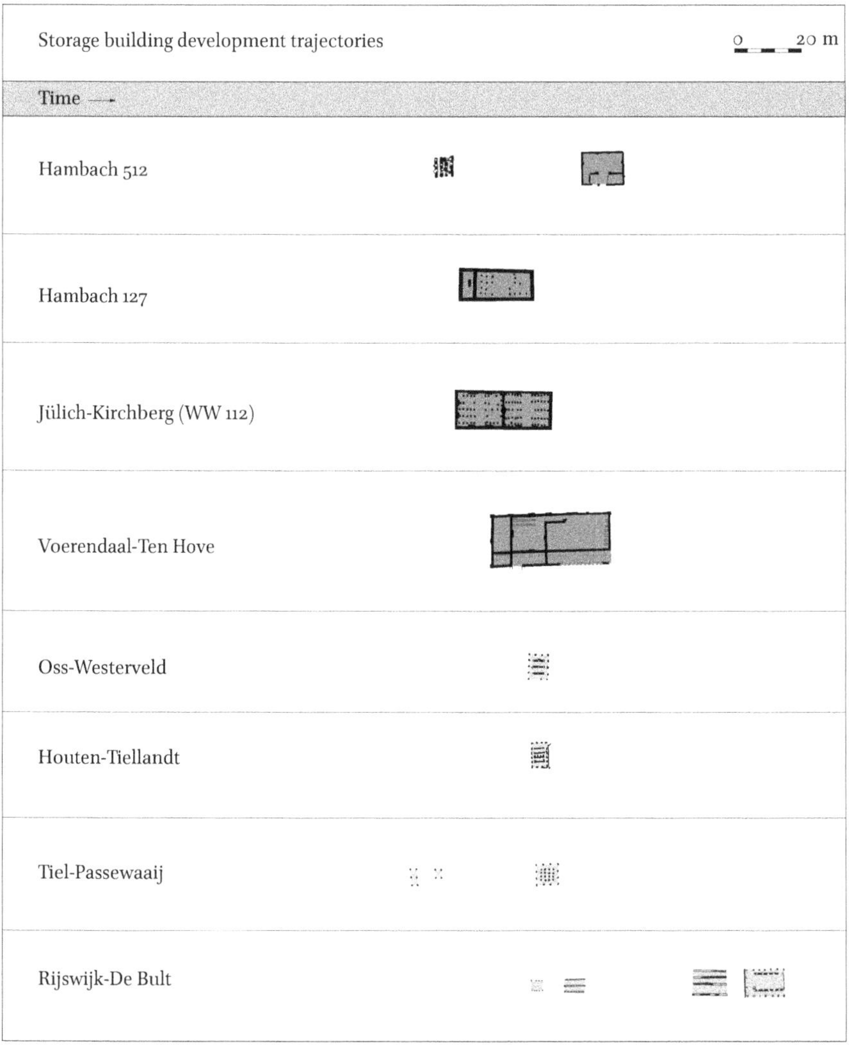

FIGURE 7.5 Évolution des *horrea* ruraux en Gaule du nord
D'APRÈS HABERMEHL 2013, 149–50 FIG. 5

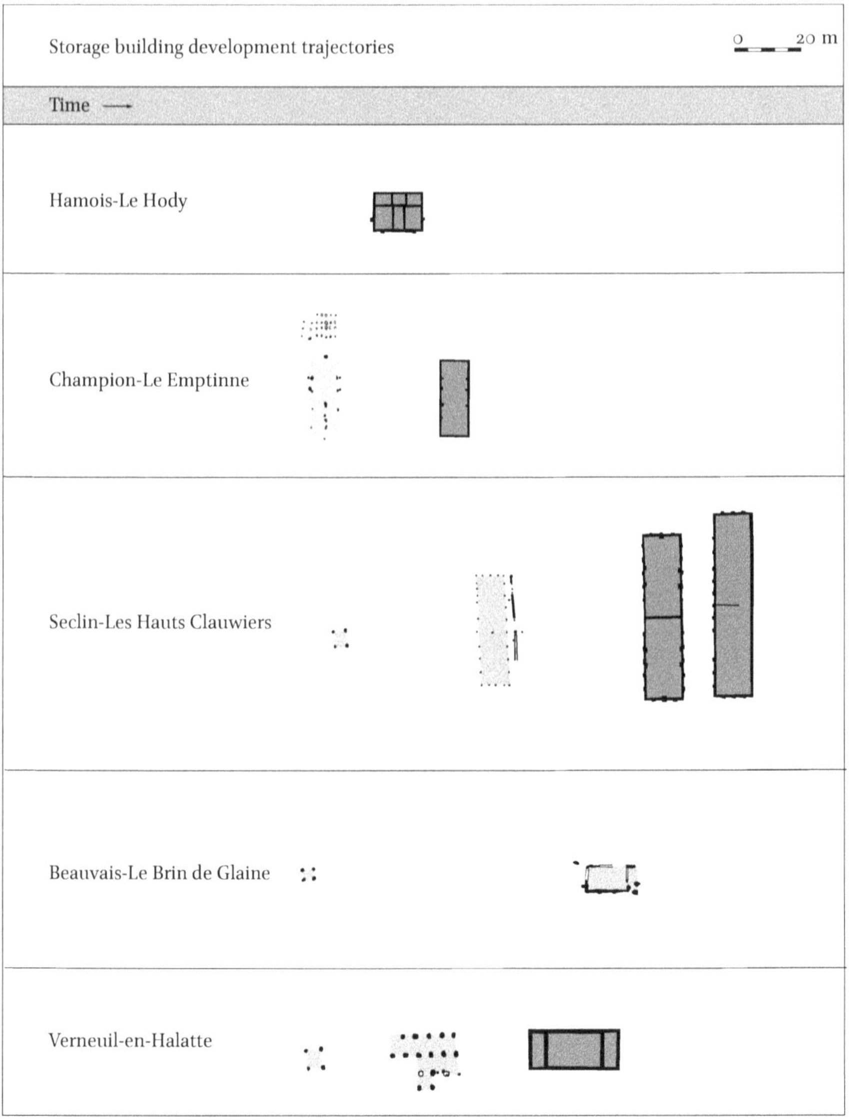

FIGURE 7.5 (cont.)

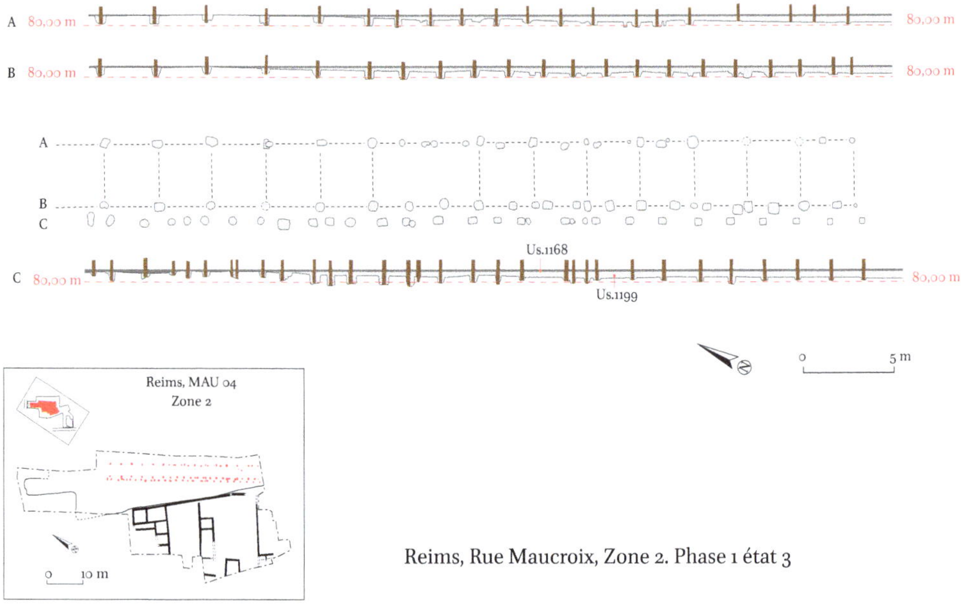

FIGURE 7.6 Les *horrea* de la rue Maucroix à Reims (état 1) D'APRÈS ROLLET ET AL. 2011

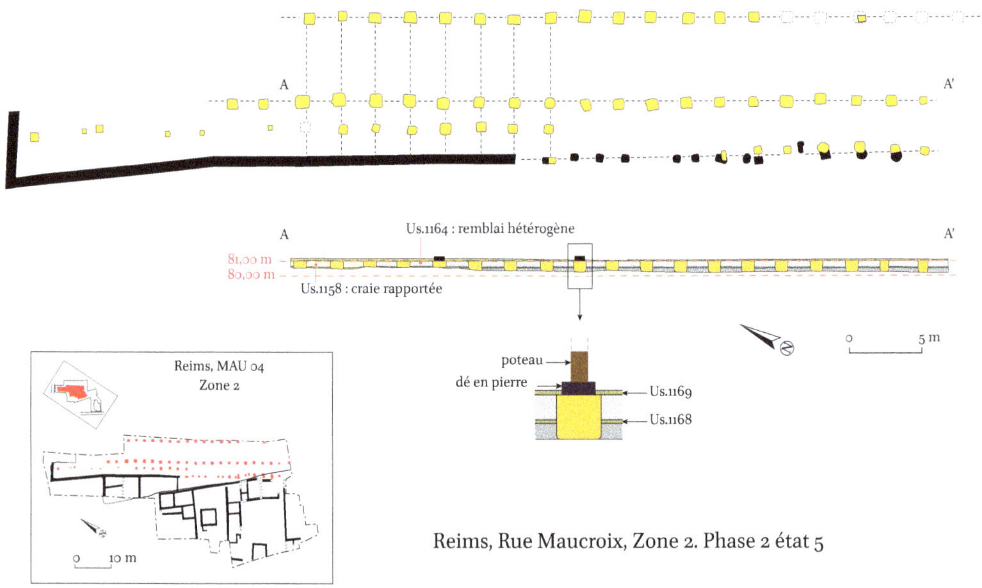

FIGURE 7.7 Les *horrea* de la rue Maucroix à Reims (état 2) D'APRÈS ROLLET ET AL. 2011

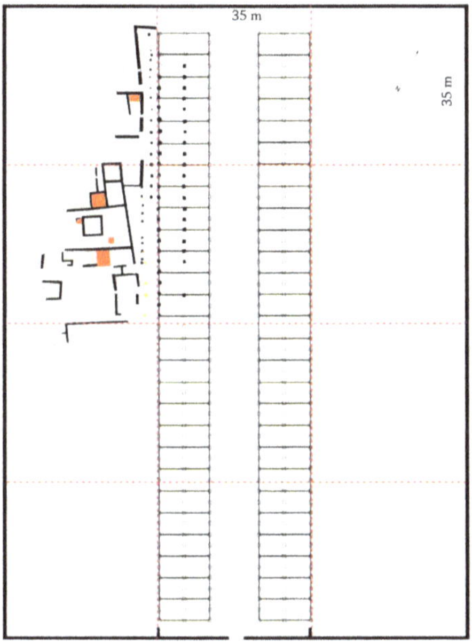

FIGURE 7.8
Les *horrea* de la rue Maucroix à Reims (état restitué)
D'APRÈS POIRIER 2011, 132

FIGURE 7.9
Les *horrea* d'Amiens, îlot de la Boucherie
D'APRÈS BINET ET PORTZAMPAC 2007, AVEC MODIFICATIONS

bout de la ville, le petit édifice mis au jour dans la "ZAC Cathédrale" a de bonnes chances, vu sa taille (90 m^2) d'être un édifice privé plutôt que public (Matterne, Yvinec, et Gemehl 1998).

Venons-en maintenant, avant de conclure, au "*granarium*" fouillé par M. Poux et son équipe à Panossas dans l'Isère (Poux 2017). La surface de stockage au sol, une fois soustraite celle du hall central, est d'environ 350 m^2, une superficie respectable, certes, mais qui n'est pas très différente de celle de Heitersheim (cat. 254), de Voerendaal ou de Alle (état 2), et paraît sensiblement moindre que celle de Biberist, de Alle (état 3), ou de Bad Rappenau. La liste pourrait être allongée. M. Poux propose une datation entre, au plus tôt, le début de l'époque flavienne et le milieu du IIIe s.

À partir de cet exemple récemment fouillé, les auteurs s'appuient sur une carte des découvertes connues de grands *horrea* pour tenter de montrer l'existence d'un réseau de collecte annonaire, lié aux grands axes du réseau routier de Gaule, à destination des villes et du *limes*, et ils en publient une carte. Si l'intention est louable et l'hypothèse recevable a priori, la démonstration pêche par l'absence de données chronologiques car le corpus mêle des données de toute nature et de toute époque. On voit ainsi figurer sur la même carte Haltern et Niederbieber, que séparent près de deux siècles. Peut-on, dès lors parler de réseau ? Sans doute de tels systèmes ont-ils bien existé, comme l'a rappelé Marie-Brigitte Carre (Carre 2011; Carre et Laudani 2016). Mais, dans le cas qui nous occupe ici, seule une enquête rigoureuse, avec des séries chronologiquement homogènes et bien datées, permettrait, peut-être, d'asseoir l'hypothèse proposée. Pour l'heure, il paraît sage de ne pas tenter de reconstituer physiquement un tel réseau d'approvisionnement.

Quelques observations s'imposent toutefois au terme de cette enquête dont la prétention n'était certes pas de conclure un débat, mais au contraire de ne pas le clore trop tôt et d'en rappeler les difficultés, d'en mieux cerner les obstacles méthodologiques.

On doit d'abord souligner l'énormité de la demande militaire à l'époque augustéenne, au moment des tentatives successives de conquête de la Germanie. Cet effort exigé des campagnes gauloises a certainement été d'autant plus important que le monde rural protohistorique, riche malgré l'impact inévitable de la guerre des Gaules encore récente, ne pouvait probablement pas fournir instantanément beaucoup plus qu'il ne produisait déjà : à population quasi stable, à conditions techniques identiques, et alors que nous ne voyons pas apparaître, à cette époque, une économie domaniale de type italien capable de fournir un surplus significatif, la demande ne pouvait à elle seule susciter l'offre de manière spontanée (Reddé 2018b). Cette idée que les nouveaux marchés militaires de Germanie suffisaient à provoquer le décollage économique de la Gaule me paraît, personnellement, parfaitement anachronique. Il est donc clair qu'on doit ici se souvenir de la fameuse phrase de Tacite (*Agr.* 19.4–5) appliquée, il est vrai, à la Bretagne flavienne : "Par une équitable répartition des charges, il rendait moins dures les prestations de blé et la levée des impôts, en retranchant ces inventions de l'avidité qu'on avait plus de peine à supporter que l'impôt lui-même. Car, par dérision, les Bretons étaient forcés d'attendre près des greniers fermés, et, qui plus est, d'acheter du blé à un prix majoré. Des routes détournées, des régions lointaines étaient assignées aux cités très voisines des quartiers d'hiver, pour porter leurs blés dans d'autres cantonnements écartés et peu accessibles, jusqu'à ce que l'arrangement le plus simple pour tous devînt pour un petit nombre une source de profit"[5]. Déjà, sous Tibère, les excès des réquisitions avaient conduit à la révolte des Gaules (Reddé 2011b). On voit bien, en comparant la capacité d'un grenier militaire de cette époque avec celle d'une ferme protohistorique, l'énorme différence de capacité qui existe entre ces deux extrémités de la chaîne de ravitaillement.

Que les campagnes gauloises aient progressivement augmenté leur potentiel avec le temps est par ailleurs une évidence. On le voit bien à travers l'évolution de la capacité de stockage des établissements agricoles, mais celle-ci n'est guère visible dans le nord avant, au plus tôt, le dernier tiers du Ier s. de n. è. Les quelques exemples qui montrent des phases successives, comme c'est le cas à Alle, attestent formellement cette progression. On ne saurait pourtant tirer de celle-ci, sans contre-épreuve, l'idée simpliste d'une augmentation équivalente de la production sur un domaine de superficie identique. Derrière l'apparition de ces *horrea* à grande capacité peuvent en effet se cacher des phénomènes de concentration foncière qui nous échappent totalement. Quant à l'idée de regroupements de stocks destinés au ravitaillement militaire, elle est parfaitement pertinente, notamment quand il s'agit de l'hinterland immédiat du *limes*, comme c'est le cas de Bad Rappenau, tout près du Neckar. Encore faudrait-il démontrer cette hypothèse de manière précise à travers une enquête plus globale car nous en sommes, pour l'instant, réduits aux spéculations.

Pour finir, les capacités de stockage devraient, dans l'absolu, être évaluées à l'échelle d'un terroir beaucoup plus qu'à celle d'un établissement considéré individuellement, comme le montre l'exemple, pour l'heure exceptionnel, du site protohistorique d'Entrammes, capable de fournir une superficie deux fois plus grande que celle du très grand *granarium* de Panossas et d'alimenter un camp légionnaire comme le "Koenenlager" de Neuss.... Idée utopique, évidemment, puisque l'archéologie n'a pas les moyens d'évaluer un terroir dans sa totalité.

---

5 Traduction Les Belles Lettres : *Frumenti et tributorum exactionem aequalitate munerum mollire, circumcisis quae in quaestum reperta ipso tributo grauius tolerabantur ; namque per ludibrium adsidere clausis horreis et emere ultro frumenta ac luere pretio cogebantur. Diuortia itinerum et longinquitas regionum indicebantur, ut ciuitates proximis hibernis in remota et auia deferrent, donec quod omnibus in promptu erat paucis lucrosum fieret.*

TABLEAU 7.1 Capacités de stockage de grain estimées pour différents greniers urbains, militaires et ruraux.

<table>
<tr><th></th><th>Lieu</th><th>Superficie au sol</th><th>Capacité estimée</th><th>Auteur de l'estimation</th><th>Commentaires</th></tr>
<tr><td rowspan="8">Sites militaires</td><td>Anreppen, Grand horreum à cour</td><td>3808 m^2</td><td></td><td rowspan="4">Kehne 2008</td><td>Pas de blé mais autres produits</td></tr>
<tr><td>Anreppen, “Horreastadt”</td><td>2369,63 m^2</td><td>1640/1968 m^3</td><td>Sacs stockés sur 1,5 m de haut (1 étage)</td></tr>
<tr><td>Anreppen Total</td><td>6177,63m^2</td><td></td><td>54 jours de rations de blé pour expédition de 25000 h en 4 apr. (1 étage)</td></tr>
<tr><td>Rödgen</td><td>3457 m^2</td><td>4149 m^3 dont 3445 pour céréales</td><td>Sacs stockés sur 1,5 m de haut = 7079 rations annuelles, environ 6 mois de réserve pour trois légions (campagnes de Drusus, 11–9 av. n. è.) (1 étage)</td></tr>
<tr><td>Neuss “Koenenlager”</td><td>858 m^2</td><td></td><td rowspan="2">Dimensions calculées d'après Petrikovits 1975</td><td></td></tr>
<tr><td>Saalburg</td><td>489 m^2</td><td></td><td></td></tr>
<tr><td rowspan="2">Walheim II</td><td>Bât. G : 452 m^2</td><td rowspan="2"></td><td rowspan="2">Reddé 2011a</td><td rowspan="2">Milieu 2e siècle. Les bâtiments G et H se succèdent.</td></tr>
<tr><td>Bât. F1: 378/400 m^2</td></tr>
<tr><td rowspan="4">Sites ruraux</td><td rowspan="2">Voerendaal</td><td rowspan="2">375 m^2</td><td>208,2/268,72 m^3</td><td>Blöck 2011–2012</td><td>Stockage en vrac, dans caissons, sur 0,55/0,70 m de hauteur (1 étage). Évaluation d'après greniers $17^{e}/18^{e}$ s.</td></tr>
<tr><td>300/400 m^3</td><td>Willems 1988, 8–13</td><td>Stockage en vrac</td></tr>
<tr><td rowspan="2">Biberist</td><td rowspan="2">400 à 590 m^2</td><td>327,57/422,78 m^3</td><td>Blöck 2011–2012</td><td>Stockage en vrac, dans caissons, sur 0,55/0,70 m de hauteur (1 étage)</td></tr>
<tr><td>1600 à 2000 m^3</td><td>Schucany 2006</td><td>Stockage en caissons, sur 4 m de hauteur (2 étages)</td></tr>
</table>

ABLEAU 7.1 Capacités de stockage de grain estimées pour différents greniers urbains, militaires (*cont.*)

	Lieu	Superficie au sol	Capacité estimée	Auteur de l'estimation	Commentaires
Sites ruraux	Heitersheim C	180 m^2	82,08/105,94 m^3	Blöck 2011–2012	Construit avant 100
	Heitersheim C2	360 m^2	158,79/204,94 m^3		Vers 120
	Panossas *	350 m^2	120 m^3	Poux 2017	Stockage en vrac sur 0,30 m (1 étage). Evaluation minimale
	Alle phase 1	48 m^2		Demarez et al. 2010	Avant 50/75 de n. è.
	Alle phase 2	392 m^2			Fin 1er s./ milieu 2e s.
	Alle phase 3	710 m^2/ surface de stockage réelle de 240 m^2	240 m^3		Avant 250/275 de n. è. Calcul de la surface de stockage réelle tenant compte des circulations internes entre les caissons avec tas de 1 m au maximum
	Bad Rappenau "Maueräcker" 2e période	608 m^2		Wulmeier et Hartmann 2009	
	Hambach 111	37,5 m^2	4500–6920 *modii*	Schubert 2016	Stockage en sacs supposé **
	Hambach 127	112,5 m^2	13840–20760 *modii*		Stockage en sacs supposé
	Hambach 130	85 m^2	10188 -15688 *modii*		Stockage en sacs supposé
	Hambach 206	58 m^2	7020–10800 *modii*		Stockage en sacs supposé
	Hambach 133	175 m^2	21000–32000 *modii*		Stockage en sacs supposé
	Hambach 303	138 m^2	16632–25584 *modii*		Stockage en sacs supposé

TABLEAU 7.1 Capacités de stockage de grain estimées pour différents greniers urbains, militaires (*cont.*)

	Lieu	Superficie au sol	Capacité estimée	Auteur de l'estimation	Commentaires
Sites urbains	Reims, rue Maucroix	50 m × 3,50 = 175 m^2 fouillés (1er état)		Reddé 2011a	Début 1er s. de n. è. (phase 1) ; entre 65–90 et 90–105 (état 2). Fouille incomplète. Restitution de la superficie dépend de la restitution. Celle qui est proposée pour l'état 2 implique une superficie totale de 2 halles de 140 m × 11,6 m soit une emprise au sol de 3248 m^2.
		Au moins 50 m × 5, 80 m = 290 m^2 (2e état)			
	Amiens, ZAC Cathédrale	20 m × 4,5 m = 90 m^2		Matterne, Yvinec et Gemehl 1998	2e s. Grenier privé ?
	Amiens, îlot Boucherie	3000 m^2 ?		Binet et Portzampac 2007	Fin 1er s. de n. è.
		îlots de 35 m × 10 m			
	Trèves	2 halles de 70 × 20 m = 2800 m^2		Merten 2005	Vers 300

* Panossas est ici classé parmi les sites ruraux, bien que M. Poux semble désormais considérer qu'il puisse s'agir d'un autre type de site.

** L'auteur compte un *modius* de 8,7 litres ou 8,7 dm^3. 4500 *modii* font donc 39150 l ou 39,15 m^3

Bibliography

Agache, Roger. 1978. *La Somme pré-romaine et romaine, d'après les prospections aériennes à basse altitude*. Mémoires de la Société des antiquaires de Picardie 24. Amiens.

ANR Entrepôts. 2012. 'Présentation du projet – contexte et enjeux'. *ANR Entrepôts*. September 28. <http://www.entrepots-anr.efa.gr/p-contexte-enjeux_fr.htm>.

Applebaum, Shimon. 1972. 'Roman Britain'. In *The agrarian history of England and Wales. Tome 1, Volume 2. A.D. 43–1042*, edited by Herbert Patrick Reginald Finberg, 1–277. Cambridge: Cambridge University Press.

Arce, Javier, and Bertrand Goffaux, eds. 2011. Horrea *d'Hispanie et de la Méditerranée romaine*. Collection de la Casa de Velázquez 125. Madrid: Casa de Velázquez.

'A Review of Methods for Estimating Grain Post-Harvest Losses'. 2015. Working paper no. 2. Rome: Global Office of the Global Strategy to improve agricultural and rural statistics. Online: <http://gsars.org/wp-content/uploads/2015/09/WP-Review-of-Methods_for_estimating_grain-loss-160915.pdf>.

Arnoldussen, Stijn, and Richard Jansen. 2010. 'Iron Age habitation patterns on the southern and northern Dutch Pleistocene coversand soils: The process of settlement nucleation'. In *Haus – Gehöft – Weiler – Dorf: Siedlungen der Vorrömischen Eisenzeit im nördlichen Mitteleuropa*, edited by Michael Meyer, 379–97. Berliner archäologische Forschungen 8. Rahden: M. Leidorf.

Asal, Markus. 2005. *Ein spätrömischer Getreidespeicher am Rhein. Die Grabung Rheinfelden-Augarten West 2001*. Veröffentlichungen der Gesellschaft Pro Vindonissa 19. Brugg: Gesellschaft pro Vindonissa.

Aulakh, Jaspreet, and Anita Regmi. 2013. 'Post-Harvest Food Losses Estimation – Development of Consistent Methodology'. Rome: FAO. Online: <http://www.fao.org/fileadmin/templates/ess/documents/meetings_and_workshops/GS_SAC_2013/Improving_methods_for_estimating_post_harvest_losses/Final_PHLs_Estimation_6-13-13.pdf>.

Auxiette, Ginette, Muriel Boulen, Sophie Desenne, Véronique Matterne, Céline Rocq, Jean-Hervé Yvinec, Philippe Ponel, and Jean-Marie Pernaud. 2003. 'Un site du Hallstatt à Villeneuve-Saint-Germain "Les Étomelles" (Aisne)'. *Revue archéologique de Picardie* 3–4: 21–65. doi:10.3406/pica.2003.2379.

Bakels, Corrie C. 1996. 'Growing Grain for Others or How to Detect Surplus Production?' *Journal of European Archaeology* 4 (1):329–36. doi:10.1179/096576696800688187.

Barral, Jean-Augustin, and Henry Sagnier. 1886. *Dictionnaire d'agriculture : encyclopédie agricole complète*. 4 vols. Paris: Hachette et Cie.

Barral, Philippe, Patrice Beck, José Bernal, François Boyer, Olivier Buchsenschutz, Jean-Loup Flouest, Joszef Laszlovsky, et al. 1995. 'Les fouilles du Mont-Beuvray, Nièvre-

Saône-et-Loire : rapport biennal 1992–1993'. *Revue Archéologique de l'Est et du Centre-Est* 46 (2): 217–93.

Barral, Philippe, Patrice Beck, and Pierre-Paul Bonenfant. 1993. 'Les fouilles du Mont-Beuvray, Nièvre-Saône-et-Loire : rapport biennal 1990–1991'. *Revue Archéologique de l'Est et du Centre-Est* 44 (1): 311–63.

Bastard, Véronique, and Anne Philippa Stephenson. 2002. 'Entre structures agraires, sanctuaire et station routière : le site de Bourbousson 3 à Crest (Drôme)'. In *Archéologie du TGV Méditerranée: fiches de synthèse 3. Antiquité, Moyen âge, époque moderne*, edited by Thierry Janin, 687–98. Monographies d'archéologie méditerranéenne 10. Lattes: Association pour la recherche archéologique en Languedoc oriental.

Bats, Adeline, ed. 2017. *Les céréales dans le monde antique. Regards croisés sur les stratégies de gestion des cultures, de leur stockage et de leur mode de consommation*. NeHeT. Revue numérique d'égyptologie 5. Paris / Bruxelles: Université Paris-Sorbonne / Université Libre de Bruxelles, online: <http://sfe-egyptologie.website/index.php?option=com_content&view=article&id=112:revue-nehet-5&catid=2:uncategorised&Itemid=234>.

Bayard, Didier, and Patrick Lemaire. 2014. 'Les vestiges de l'Antiquité sur les grands tracés linéaires en Picardie'. *Revue Archéologique de Picardie* 3–4: 7–232.

Beaujard, Stéphane, François Malrain, and Ginette Auxiette. 2006. 'Allonne : une ferme laténienne au cœur du territoire bellovaque'. *Revue archéologique de Picardie* 3–4: 5–20. doi:10.3406/pica.2006.2432.

Bechert, Tilmann. 1978. 'Wachtturm oder Kornspeicher? Zur Bauweise spätrömischer Burgi'. *Archäologisches Korrespondenzblatt* 8: 127–32.

Bénard, Jacky, and Michel Mangin. 1990. 'Alésia. L'agglomération gauloise'. In *Archéologie de Bourgogne 1. La Côte-d'Or*, 43–45. Dijon.

Bernardos Sanz, José Ubaldo, and Catherine Virlouvet. 2016. 'Formes et fonctions : essai de typologie'. In *Entrepôts et trafics annonaires et Méditerranée. Antiquité – Temps modernes*, edited by Brigitte Marin and Catherine Virlouvet, 59–82. Collection de l'École française de Rome 522. Rome: École française de Rome.

Besnard-Vauterin, Chris-Cécile, ed. 2009. *En plaine de Caen, une campagne gauloise et antique*. Archéologie & culture. Rennes: Presses universitaires de Rennes.

Besnard-Vauterin, Chris-Cécile, ed. 2011. *Ifs – ZAC Object'Ifs Sud – Dernière Tranche – Calvados. Habitats et Lieux Funéraires Protohistoriques et Vestiges Antiques*. Rapport de fouille d'archéologie préventive. Caen: SRA Normandie / Inrap.

Beutler, Corinne. 1981. 'Traditions et innovations dans les techniques de conservation des grains à la campagne et à la ville (Europe occidentale, xvème-xviiième siècle). Examen critique des sources'. In *Les techniques de conservation des grains à long terme. Leur rôle dans la dynamique des systèmes de cultures et des sociétés 2*, edited by Marceau Gast, François Sigaut, and Ariane Bruneton-Governatori, 19–43. Paris: Éditions du CNRS.

Binet, Éric. 2007. 'Amiens – Îlot de la Boucherie'. *ADLFI. Archéologie de la France – Informations. une revue Gallia*, online : <https://journals.openedition.org/adlfi/5588>.

Binet, Éric, and Pierre de Portzampac. 2007. 'Des entrepôts et un théâtre gallo-romains à Amiens [dépliant de site]'. INRAP, online: <http://www.images-archeologie.fr/Accueil/Recherche/p-12-notice-IMPRIME-Des-entrepots-et-un-theatre-gallo-romains-a-Amiens.htm?&notice_id=1719>.

Black, E.W. 1981. 'An Additional Classification of Granaries in Roman Britain'. *Britannia* 12: 163–65. doi:10.2307/526247.

Blancquaert, Geertrui, and Yves-Marie Adrian. 2006. 'Les occupations multiples de la Plaine du Bosc Renault à Hautot-le-Vatois (Seine-Maritime) : la zone de stockage du premier âge du Fer et les vestiges antiques funéraires et domestiques'. *Revue archéologique de l'Ouest* 23: 9–44. doi:10.4000/rao.75.

Blancquaert, Geertrui, Chantal Leroyer, Thierry Lorho, François Malrain, and Véronique Zech-Matterne. 2012. 'Rythmes de créations et d'abandons des établissements ruraux du second âge du Fer et interactions environnementales'. In *Variabilités environnementales, mutations sociales. Nature, intensités, échelles et temporalités des changements. Actes des rencontres, 20–22 octobre 2011*, edited by Frédérique Bertoncello and Frank Braemer, 233–45. Rencontres internationales d'archéologie et d'histoire d'Antibes 32. Antibes: Éd. APDCA.

Blancquaert, Geertrui, Thierry Lorho, François Malrain, and Yves Menez. 2009. 'Bilan et perspectives de recherches sur les sites ruraux au second Âge du Fer'. In *Les Gaulois entre Loire et Dordogne. Actes du XXXIe colloque international de l'AFEAF, 17–20 mai 2007, Chauvigny (Vienne, F)*, edited by Isabelle Bertrand, Alain Duval, and José Gomez de Soto, 5–23. Mémoire – Société de recherches archéologiques de Chauvigny 35. Chauvigny: Association des Publications Chauvinoises.

Blancquaert, Geertrui, and François Malrain, eds. 2016. *Évolution des sociétés gauloises du second âge du Fer, entre mutations internes et influences externes. Actes du 38e colloque international de l'AFEAF, Amiens, 29 mai–1er juin 2014*. Revue Archéologique de Picardie. Numéro Spécial 30. Amiens: Revue archéologique de Picardie.

Blöck, Lars. 2008. 'Ein Speichergebäude und spätlatènezeitliche Funde in der römischen Villa im Ortskern von Grenzach, Gde. Grenzach-Wyhlen, Kreis Lörrach'. *Archäologische Ausgrabungen in Baden-Württemberg*, 156–60.

Blöck, Lars. 2011–2012. 'Die Erweiterung der Getreidespeicherkapazitäten der Axialhofvilla Heitersheim in ihrer 4. Bauperiode – Binnenkolonisation oder Konzentrationsprozesse im ländlichen Raum im ausgehenden 2. Jahrhundert n. Chr.? Ein Modell zur Berechnung von Getreideanbauflächen anhand der Speicherkapazität römischer Horrea'. *Alemannisches Jahrbuch* 59/60: 81–111.

Blöck, Lars. 2016. *Die Römerzeitliche Besiedlung im rechten südlichen Oberrheingebiet.* Forschungen und Berichte zur Archäologie in Baden-Württemberg 1. Wiesbaden: Dr. Ludwig Reichert Verlag.

Bloemers, Johan Hendrik Frederik. 1978. *Rijswijk (Z. H.), 'De Bult'. Eine Siedlung der Cananefaten.* 3 vols. Nederlandse oudheden 8. Amersfoort: ROB.

Boelcke, Willi Alfred. 1964. 'Bäuerlicher Wohlstand in Württemberg Ende des 16. Jahrhunderts'. *Jahrbücher für Nationalökonomie und Statistik* 176: 241–280. doi:10.1515/jbnst-1964-0114.

Boetto, Giulia, Évelyne Bukowiecki, Nicolas Monteix, and Corinne Rousse. 2016. 'Les *Grandi Horrea* d'Ostie'. In *Entrepôts et trafics annonaires et Méditerranée. Antiquité – Temps modernes*, edited by Brigitte Marin and Catherine Virlouvet, 177–226. Collection de l'École française de Rome 522. Rome: École française de Rome. <https://halshs.archives-ouvertes.fr/halshs-01441802>.

Bonacci, Lucia. 2015. 'Il grano gallico: magazzini di stoccaggio e tempi di trasporto a Roma'. *Pallas. Revue d'études antiques* 99: 175–91. doi:10.4000/pallas.3083.

Borgongino, Michele. 2006. *Archeobotanica. Reperti vegetali da Pompei e dal territorio vesuviano.* Studi della Soprintendenza archeologica di Pompei 16. Rome: L'Erma di Bretschneider.

Booth, Paul M. 1989. 'Roman Store Buildings in Alcester'. *Transactions of the Birmingham and Warwickshire Archaeological Society* 94: 63–106.

Bossard, Stanislas. 2015. 'Les souterrains et autres architectures enterrées de l'âge du Fer en Bretagne et Basse-Normandie. Analyse de structures de stockage spécifiques au nord-ouest de la Gaule (VIe–Ier siècle avant n. è.)'. Mémoire de Master 2, Nantes: Université de Nantes.

Bossard, Stanislas, Gérard Guillier, Hubert Lepaumier, and Axel Levillayer. 2018. 'Les architectures des structures de stockage alimentaire à l'âge du Fer dans l'ouest de la Gaule (VIe–Ier s. av. n. è.) : choix techniques ou culturels ?' In *Architectures de l'âge du Fer en Europe occidentale et centrale. Actes du 40e colloque de l'AFEAF (4–7 mai 2016, Rennes)* edited by Anne Villard-Le Tiec: 349–88. Rennes: Presses Universitaires de Rennes.

Bouby, Laurent. 2001. 'L'orge à deux rangs (*Hordeum distichum*) dans l'agriculture gallo-romaine : données archéobotaniques'. *Revue d'Archéométrie* 25 (1):35–44. doi: 10.3406/arsci.2001.999.

Bowman, Alan Keir, and Andrew Wilson. 2013a. 'Introduction: Quantifying Roman Agriculture'. In *The Roman agricultural economy. Organisation, investment, and production*, edited by Alan Keir Bowman and Andrew Wilson, 1–32. Oxford Studies on the Roman Economy 3. Oxford: Oxford University Press.

Bowman, Alan Keir, and Andrew Wilson, eds. 2013b. *The Roman Agricultural Economy. Organisation, Investment, and Production.* Oxford Studies on the Roman Economy 3. Oxford: Oxford University Press.

Brinkkemper, Otto. 1993. *Wetland Farming in the Area to the South of the Meuse Estuary during the Iron Age and Roman Period. An Environmental and Palaeo-Economic Reconstruction.* Analecta Praehistorica Leidensia 24. Leiden: Leiden University Press.

Brinkkemper, Otto, H. Duistermaat, D.P. Hallewas, and Laura I. Kooistra. 1995. 'A Native Settlement from the Roman Period near Rockanje. Archaeology, Environmental Reconstruction, and Economy'. *Berichten van de Rijksdienst voor het Oudheidkundig Bodemonderzoek* 41: 123–71.

Brück, Joachim. 1989. 'Die Ausgrabung einer römischen Grossvillenanlage bei Borg'. *Saarheimat* 33: 196–99.

Brück, Joachim. 1992. 'Die Villa von Borg'. In *Der Kreis Merzig-Wadern und die Mosel Zwischen Nenning und Metz*, edited by Jan Lichardus and Andrei Miron. Führer zu Archäologischen Denkmälern in Deutschland 24. Stuttgart: Thiess.

Brulet, Raymond, ed. 2008. *Les Romains en Wallonie.* Bruxelles: Racine.

Buckland, Paul. 1978. 'Cereal production, storage, and population: a caveat'. In *The Effect of man on the landscape: the Lowland zone*, edited by Susan Limbrey and John Gwynne Evans, 43–45. Council for British Archaeology Research Report 21. London: Council for British Archaeology. <http://archaeologydataservice.ac.uk/archives/view/cba_rr/rr21.cfm>.

Busana, Maria Stella. 2002. *Architetture rurali nella Venetia romana.* Le rovine circolari 3. Roma: L'Erma di Bretschneider.

Carre, Marie-Brigitte. 2011. 'Les réseaux d'entrepôts dans le monde romain. Étude de cas'. In *Horrea d'Hispanie et de la Méditerranée romaine*, edited by Javier Arce and Bertrand Goffaux, 23–39. Collection de la Casa de Velázquez 125. Madrid: Casa de Velázquez.

Carre, Marie-Brigitte, and Simona Laudani. 2016. 'Distribution géographique des entrepôts, localisations, réseaux : étude de cas'. In *Entrepôts et trafics annonaires et Méditerranée. Antiquité – Temps modernes*, edited by Brigitte Marin and Catherine Virlouvet, 13–57. Collection de l'École française de Rome 522. Rome: École française de Rome.

Censier, Damien. 2013. *Cysoing. Clos de l'Abbaye. Nord-Pas-de-Calais / Nord (59).* 4 vols. Rapport Final d'opération. Douai: SRA Nord-Pas-de-Calais / Communauté d'agglomération du Douaisis, Direction de l'archéologie preventive.

Chancrin, Ernest, and René Dumont, eds. 1921. *Larousse agricole. Encyclopédie illustrée.* Paris, Larousse.

Chankowski, Véronique, Xavier Lafon, and Catherine Virlouvet, eds. 2018. *Entrepôts et circuits de distribution en Méditerranée antique.* Bulletin de Correspondance Hellénique. Supplément 58. Athènes: École française d'Athènes.

Charrondiere, U. Ruth, David Haytowitz, and Barbara Stadlmayr. 2015. 'Base de données FAO/INFOODS sur la densité. Version 2.0 (2015)'. FAO, online: <http://www.fao.org/3/a-i3057f.pdf>.

Chaume, Bruno, Alfred Haffner, Norbert Niezery, and Walter Reinhard. 2011. 'L'organisation spatiale du plateau Saint-Marcel : un habitat urbanisé de la fin du premier âge du Fer'. In *Le complexe aristocratique de Vix. Nouvelles recherches sur l'habitat, le système de fortification et l'environnement du mont Lassois*, edited by Bruno Chaume and Claude Mordant, 367–81. Art, Archéologie et Patrimoine. Dijon: Éd. universitaires de Dijon.

Coulon, Gérard. 1987. 'L'implantation gallo-romaine en Bas-Berry. Présentation de quelques découvertes récentes et de sites méconnus'. *Revue de l'Académie du Centre* 1987: 39–58.

Cuvillier, Jean. 1931. 'Conservation des récoltes'. In *Encyclopédie agricole Quillet*, by Auguste Sartory and Georges Couturier, 341–99. Paris: A. Quillet.

Da Gai, Enrico, and Giulia Vertecchi. 2016. 'Un document inédit sur le calcul des capacités de stockage des entrepôts : comparaison entre Rome et Venise'. In *Entrepôts et trafics annonaires et Méditerranée. Antiquité – Temps modernes*, edited by Brigitte Marin and Catherine Virlouvet, 313–22. Collection de l'École française de Rome 522. Rome: École française de Rome.

De Clercq, Wim. 2009. 'Lokale gemeenschappen in het Imperium Romanum: transformaties in de rurale bewoningsstructuur en de materiële cultuur in de landschappen van het noordelijk deel van de civitas Menapiorum (Provincie Gallia-Belgica, ca. 100 v. Chr. – 400 n. Chr.)'. Dissertation, Gent: Universiteit Gent. <http://hdl.handle.net/1854/LU-2096483>.

De Clercq, Wim. 2011. 'Roman rural settlements in Flanders. Perspectives on a "non-villa" landscape *in extrema Galliarum*'. In *Villa landscapes in the Roman North. Economy, culture and lifestyles*, edited by Nico Roymans and Ton Derks, 235–57. Amsterdam archaeological studies 17. Amsterdam: Amsterdam University Press.

De Lucia, M., and D. Assenato. 1992. *L'après-récolte des grains. Organisation et techniques*. Bulletin des services agricoles de la FAO 93. Roma: FAO, online : <http://www.fao.org/docrep/T0522F/T0522F00.htm>.

De Lucia, M., and D. Assenato. 1994. *Agricultural engineering in development. Post-harvest operations and management of foodgrains*. FAO agricultural services bulletin 93. Rome: FAO, online : <http://www.fao.org/docrep/t0522e/T0522E00.htm>.

De Muylder, Marjolaine. 2014. *Canal Seine-Nord Europe, fouille 34, Picardie, Oise, Noyon, 'La Mare Aux Canards' : une villa aristocratique de la cité des Viromanduens*. 9 vols. Rapport de Fouille, Canal Seine-Nord Europe, fouille 34. Amiens, Croix-Moligneaux: SRA Hauts-de-France / Inrap NP, CSNE.

De Romanis, Federico. 2007. 'In tempi di guerra e di peste. *Horrea* e mobilità del grano pubblico tra gli Antonini e i Severi'. *Antiquités africaines* 43:187–230. doi:10.3406/antaf.2007.1425.

Deffressigne, Sylvie, and Michaël Landolt. 2017. 'L'évolution du stockage entre le XIe et le IIIe s. avant notre ère dans les vallées de la Moselle, de la Meurthe et du Rhin'. In *Production et proto-industrialisation aux âges du Fer. Perspectives sociales et environnementales. Actes du 39e colloque international de l'AFEAF (Nancy, 14–17 mai 2015)*, edited by Stéphane Marion, Sylvie Deffressigne, Jenny Kaurin, and Gérard Bataille, 77–100. Mémoires – Ausonius 47. Bordeaux: Ausonius Éditions.

Deffressigne, Sylvie, Nicolas Tikonoff, Karine Boulanger-Bouchet, Christine Chaussé, and Corinne Tesnier-Hermetey. 2002. 'Les gisements d'habitat de la fin du premier âge du Fer à Gondreville – Fontenoy-sur-Moselle (54)'. *Archaeologia Mosellana* 4: 81–184.

Demarez, Jean-Daniel. 1987. *Les bâtiments à fonction économique dans les* fundi *de la* Provincia Belgica. Amphora 50. Bruxelles: Amphora.

Demarez, Jean-Daniel, Blaise Othenin-Girard, Denis Aubry, Christoph Brombacher, and Jean Detrey. 2010. *Établissements ruraux de La Tène et de l'époque romaine à Alle et Porrentruy (Jura, Suisse)*. Cahiers d'archéologie jurassienne 28. Porrentruy: Société jurassienne d'Emulation : Office du patrimoine historique. <https://doc.rero.ch/record/255029/>.

Desmoulins, A. 1896. *Procédés de conservation des produits et denrées agricoles*. Encyclopédie scientifique des aide-mémoire. Paris: Masson / Gauthier-Villars. <http://gallica.bnf.fr/ark:/12148/bpt6k2984636>.

Devauges, Jean-Bernard. 1979. 'Informations archéologiques. Circonscription de Bourgogne'. *Gallia* 37 (2): 437–68. <http://www.persee.fr/doc/galia_0016-4119_1979_num_37_2_1615>.

Diffloth, Paul. 1907. *Agriculture générale. 2, Semailles & récoltes*. Encyclopédie agricole. Paris: J.B. Baillière et fils. <https://archive.org/details/agriculturegnraoodiffgoog>.

Diffloth, Paul. 1917. *La Conservation des récoltes, grains, fourrages, racines et tubercules, pulpes, plantes industrielles, dessiccation des produits et résidus agricoles*. Paris: J.-B. Baillière et fils.

Dinter, Marieke van. 2013. 'The Roman Limes in the Netherlands: How a Delta Landscape Determined the Location of the Military Structures'. *Netherlands Journal of Geosciences* 92 (1): 11–32. doi:10.1017/S0016774600000251.

Dondin-Payre, Monique, and Marie-Thérèse Raepsaet-Charlier, eds. 1999. *Cités, municipes, colonies. Les processus de municipalisation en Gaule et en Germanie sous le Haut Empire romain*. Publications de la Sorbonne. Histoire ancienne et médiévale 53. Paris: Publications de la Sorbonne.

Dorow, Wilhelm. 1826. *Römische Alterthümer in und um Neuwied am Rhein, mit Grundrissen, Aufrissen und Durchschnitten des daselbst ausgegrabenen Kastells, und*

Darstellungen der darin gefundenen Gegenstände. Berlin: Schlesinger. <http://archive.org/details/rmischealterthoodoro>.

Dubouloz, Julien. 2008. 'Propriété et exploitation des entrepôts à Rome et en Italie (Ier–IIIe siècles)'. *Mélanges de l'École française de Rome* 120 (2): 277–94. doi:10.3406/mefr.2008.10471.

Duhamel du Monceau, Henri-Louis. 1753. *Traité de la conservation des grains et en particulier du froment*. Paris: Chez Hippolyte-Louis Guerin & Louis-François Delatour. <http://dx.doi.org/10.3931/e-rara-21144>.

Duhamel du Monceau, Henri-Louis. 1768. *Traité de la conservation des grains, et en particulier du froment. Troisième édition*. Paris: Chez Louis-François Delatour. <https://archive.org/details/bub_gb_3hwOAAAAQAAJ>.

Dureau de La Malle, Adolphe. 1840. *Économie politique des Romains*. 2 vols. Paris: Hachette. <https://archive.org/details/economiepolitiquo1dureuoft> and <https://archive.org/details/economiepolitiquoodure>.

Durliat, Jean. 1990. *De la ville antique à la ville byzantine. Le problème des subsistances*. Collection de l'École française de Rome 136. Rome: École française de Rome. <http://www.persee.fr/doc/efr_0000-0000_1990_ths_136_1>.

Echt, Rudolf, Bettina Birkenhagen, and Florian Sărăţeanu-Müller, eds. 2016. *Monumente der Macht. Die gallo-römischen Großvillen vom längsaxialen Typ. Internationale Tagung vom 26. bis 28. März 2009 im Archäologiepark Römische Villa Borg*. Saarbrücker Beiträge zur Altertumskunde 90. Bonn: Rudolf Habelt Gmbh.

Eck, Werner. 2007. 'Das römische Köln: Wie deckt eine Provinzstadt ihren Bedarf?' In *Supplying Rome and the Empire. Proceedings of an international seminar held at Siena-Certosa di Pontignano on May 2–4, 2004, on Rome, the provinces, production and distribution*, edited by Emanuele Papi, 209–18. Journal of Roman archaeology. Supplementary series 69. Portsmouth: Journal of Roman archaeology.

Enckevort, Harry van. 2012. *Gebundelde sporen. Enkele kanttekeningen bij aardewerk en nederzettingen uit Romeins Nederland*. Archeologische Berichten Nijmegen 7. Leiden: Universiteit Leiden. <https://openaccess.leidenuniv.nl/handle/1887/20151>.

Erdkamp, Paul. 2005. *The grain market in the Roman Empire. A social, political and economic study*. Cambridge: Cambridge University Press.

Faoro, Davide. 2015. 'Die Grossprovinz Germanien: idea e forma di macroprovincia augustea'. In *Il* princeps *romano: autocrate o magistrato? Fattori giuridici e fattori sociali del potere imperiale da Augusto a Commodo*, edited by Jean-Louis Ferrary and John Scheid, 679–700. Pubblicazioni del CEDANT 14. Pavia: IUSS Press.

Ferdière, Alain. 1985. 'Recherches sur les contextes de découvertes archéologiques des céréales'. In *Les techniques de conservation des grains à long terme. Leur rôle dans la dynamique des systèmes de cultures et des sociétés 3*, edited by Marceau Gast and François Sigaut, 2:357–75. Paris: Éd. du CNRS.

Ferdière, Alain. 1988a. *Les campagnes en Gaule romaine (52 av. J.-C.-486 ap. J.-C.). Tome 1, Les hommes et l'environnement en Gaule rurale*. Collection des Hespérides. Paris: Éd. Errance.

Ferdière, Alain. 1988b. *Les campagnes en Gaule romaine (52 av. J.-C.-486 ap. J.-C.). Tome 2, Les techniques et les productions rurales en Gaule*. Collection des Hespérides. Paris: Éd. Errance.

Ferdière, Alain. 2009. 'Chronique "Voyage à travers la Gaule profonde" – XIV'. *Revue archéologique du Centre de la France* 48: 221–29. <https://journals.openedition.org/racf/1350>.

Ferdière, Alain. 2015. 'Essai de typologie des greniers ruraux de Gaule du Nord'. *Revue archéologique du Centre de la France* 54, online: <https://journals.openedition.org/racf/2294>.

Ferdière, Alain, Cristina Gandini, and Pierre Nouvel. 2017. 'Les granges de plan carré à porche entre deux pavillons, un édifice pluri-fonctionnel emblématique de la Gaule centrale'. In *Produire, transformer et stocker dans les campagnes des Gaules romaines. Problèmes d'interprétation fonctionnelle et économique des bâtiments d'exploitation et des structures de production agro-pastorale. Colloque international AGER XI (Clermont-Ferrand, 11–13 juin 2014)*, edited by Frédéric Trément, 157–180. Aquitania. Supplément. Bordeaux: Fédération Aquitania.

Ferdière, Alain, Cristina Gandini, Pierre Nouvel, and Jean-Luc Collart. 2010. 'Les grandes *villae* "à pavillons multiples alignés" dans les provinces des Gaules et des Germanies : répartition, origine et fonctions'. *Revue archéologique de l'Est* 59 (2): 357–446. <https://journals.openedition.org/rae/6217>.

Fernández Ochoa, Carmen, Fernando Gil Sendino, Javier Salido Domínguez, and Mar Zarzalejos Prieto. 2012. *El* horreum *de la* villa *romana de Veranes (Gijón, Asturias). Primer testimonio material de los hórreos de Asturias*. Arte y Humanidades. Madrid: Universidad Nacional de Educación a Distancia.

Ferrio, Juan P., Natàlia Alonso, Jordi Voltas, and José Luis Araus. 2004. 'Estimating Grain Weight in Archaeological Cereal Crops: A Quantitative Approach for Comparison with Current Conditions'. *Journal of Archaeological Science* 31 (11): 1635–42. doi: 10.1016/j.jas.2004.04.006.

Finding the limits of the Limes. 2012. 'Full Project Proposal'. *Finding the Limits of the Limes*. July 30. <https://limeslimits.wordpress.com/full-project-proposal/>.

Fouillet, Nicolas, and Gaëlle Morillon. 2017. 'Les greniers maçonnés ruraux antiques à plancher surélevé dans les provinces des Gaules et des Germanies'. In *Produire, transformer et stocker dans les campagnes des Gaules romaines. Problèmes d'interprétation fonctionnelle et économique des bâtiments d'exploitation et des structures de production agro-pastorale. Colloque international AGER XI (Clermont-Ferrand, 11–13 juin 2014)*, edited by Frédéric Trément, 389–407. Aquitania. Supplément. Bordeaux: Fédération Aquitania.

Fremersdorf, Fritz. 1933. *Der römische Gutshof Köln-Müngersdorf*. Römisch-germanische Forschungen 6. Berlin: W. de Gruyter.

Galiani, Ferdinando. 1770. *L'Art de conserver les grains, par Barthélemy Inthiery, ouvrage traduit de l'italien par les soins de M. D. N. E., ancien officier de cavalerie*. Paris: Chez Saugrain. <http://gallica.bnf.fr/ark:/12148/bpt6k56597802>.

Gallant, Thomas W. 1991. *Risk and survival in ancient Greece. Reconstructing the rural domestic economy*. Cambridge: Polity Press.

Galsterer, Hartmut. 1990. '*Plebiculam pascere*. Die Versorgung Roms in der Kaiserzeit'. *Critica storica. Associazione degli storici europei* 27:21–40.

Garnsey, Peter. 1994. 'L'approvisionement des armées et la ville de Rome'. In *Le ravitaillement en blé de Rome et des centres urbains des débuts de la République jusqu'au Haut Empire. Actes du colloque de Naples, 14–16 février 1991*, 31–34. Collection de l'École Française de Rome 196 / Collection du Centre Jean Bérard 11. Rome / Naples: École française de Rome / Centre Jean Bérard. <http://www.persee.fr/doc/efr_0000-0000_1994_act_196_1_4651>.

Garnsey, Peter, Thomas W. Gallant, and Dominic Rathbone. 1984. 'Thessaly and the Grain Supply of Rome during the Second Century B.C.' *The Journal of Roman Studies* 74:30–44. doi:10.2307/299005.

Gast, Marceau, and François Sigaut, eds. 1979. *Les techniques de conservation des grains à long terme. Leur rôle dans la dynamique des systèmes de culture et des sociétés*. Paris: Éd. du CNRS.

Gast, Marceau, François Sigaut, Corinne Beutler, and Olivier Buchsenschutz, eds. 1985. *Les techniques de conservation des grains à long terme. Leur rôle dans la dynamique des systèmes de cultures et des sociétés 3*. 2 vols. Paris: Éd. du CNRS.

Gast, Marceau, François Sigaut, and Ariane Bruneton-Governatori, eds. 1981. *Les techniques de conservation des grains à long terme. Leur rôle dans la dynamique des systèmes de cultures et des sociétés 2*. Paris: Éditions du CNRS.

Gentry, Anne P. 1976. *Roman military stone-built granaries in Britain*. British Archaeological Reports 32. Oxford: British Archaeological Reports.

Geraci, Giovanni. 2008. 'Granai nell'Egitto ellenistico e romano: problemi tipologici, lessicali, funzionali e metodologici'. *Mélanges de l'École française de Rome* 120 (2): 307–22. doi:10.3406/mefr.2008.10473.

Geraci, Giovanni. 2008. 2015. 'Stoccagio e conservazione del grano: lo scenario antico'. *Simblos. Scritti di storia antica* 6:201–26.

Geraci, Giovanni, and Brigitte Marin. 2016. 'Stockage et techniques de conservation des grains'. In *Entrepôts et trafics annonaires et Méditerranée. Antiquité – Temps modernes*, edited by Brigitte Marin and Catherine Virlouvet, 83–136. Collection de l'École française de Rome 522. Rome: École française de Rome.

Gerritsen, Fokke. 2003. *Local identities. Landscape and community in the late prehistoric Meuse-Demer-Scheldt region.* Amsterdam archaeological studies 9. Amsterdam: Amsterdam University Press. <http://www.oapen.org/search?identifier=340244>.

Gransar, Frédéric. 2000. 'Le stockage alimentaire sur les établissements ruraux de l'âge du Fer en France septentrionale : complémentarité des structures et tendances évolutives'. In *Les installations agricoles de l'âge du Fer en France septentrionale*, edited by Gertrude Blancquaert and Stéphane Marion, 277–97. Études d'histoire et archéologie 6. Paris: Éd. Rue d'Ulm, Presses de l'École normale supérieure.

Gransar, Frédéric. 2001. 'Le stockage alimentaire à l'âge du fer en Europe tempérée'. Thèse de doctorat, Paris: Université Paris I Panthéon-Sorbonne.

Gransar, Frédéric. 2002. 'La batterie de silos de Soupir "Le Champ Grand Jacques" (Aisne) : contribution à l'identification d'une centralisation du stockage à La Tène B dans le nord de la France'. In *Découvertes récentes de l'âge du Fer dans le massif des Ardennes et ses marges. Repas des vivants et nourriture pour les morts en Gaule. Actes du XXVe Colloque International de l'AFEAF, Charleville-Mézières, 24–27 mai 2001*, edited by Patrice Méniel and Bernard Lambot, 67–80. Mémoires de la Société Archéologique Champenoise 16. Reims: Société archéologique champenoise.

Gransar, Frédéric. 2003. 'L'apport de l'étude du stockage à la reconstitution des sytèmes agro-alimentaires de l'âge du Fer en France septentrionale'. In *Le traitement des récoltes. Un regard sur la diversité du néolithique au présent*, edited by Patricia C. Anderson, Linda S. Cummings, Thomas K. Schippers, and Bernard Simonel, 201–17. Rencontres internationales d'archéologie et d'histoire d'Antibes 23. Antibes: Éd. APDCA.

Gransar, Frédéric, Véronique Matterne, and Claudine Pommepuy. 1999. 'Témoins archéologiques de la chaîne opératoire de traitement des céréales vêtues à l'âge du Fer dans le Nord de la France'. In *Els productes alimentaris d'origen vegetal a l'edat del ferro de l'Europa occidental: de la producció al consum. Actes del XXII Col·loqui Internacional per a l'Estudi de l'Edat del Ferro*, edited by Ramón Buxó and Enriqueta Pons i Brun, 237–49. Sèrie monogràfica 18. Girona: Museu d'Arqueologia de Catalunya, Generalitat de Catalunya.

Grenier, Albert. 1934. *Manuel d'archéologie gallo-romaine. 2, L'archéologie du sol. 1, Les routes*. Vol. 1. Manuel d'archéologie préhistorique, celtique et gallo-romaine 6. Paris: A. Picard.

Groenewoudt, Bert. 2011. 'The Visibility of Storage'. In *Food in the Medieval Rural Environment: Processing, Storage, Distribution of Food*, edited by Jan Klápšte and Petr Sommer, 187–97. Ruralia 8. Turnhout: Brepols Publishers.

Groot, Maaike. 2011. 'Household specialisation in horse breeding: the role of returning veterans in the Batavian river area'. In *Fines imperii – imperium sine fine ? Römische Okkupations- und Grenzpolitik im frühen Principat. Beiträge zum Kongress 'Fines imperii-imperium sine fine ?' in Osnabrück vom 14. bis 18. September 2009*, edited by

Günther Moosbauer and Rainer Wiegels, 203–18. Osnabrücker Forschungen zu Altertum und Antike-Rezeption 14. Rahden: VML, Verlag Marie Leidorf.

Groot, Maaike. 2016. *Livestock for sale: animal husbandry in a Roman frontier zone. The case study of the* Civitas Batavorum. Amsterdam archaeological studies 24. Amsterdam: Amsterdam University Press.

Groot, Maaike, Stijn Heeren, Laura I. Kooistra, and Wouter K. Vos. 2009. 'Surplus Production for the Market? The Agrarian Economy in the Non-Villa Landscapes of Germania Inferior'. *Journal of Roman Archaeology* 22 (1): 231–52. doi:10.1017/S1047759400020687

Groot, Maaike, and Laura I. Kooistra. 2009. 'Land Use and the Agrarian Economy in the Roman Dutch River Area'. *Internet Archaeology*, no. 27, online: <http://intarch.ac.uk/journal/issue27/groot_index.html>. doi:10.11141/ia.27.5.

Guesle-Coquelet, Catherine. 2011. *Les capitales de cité des provinces de Belgique et de Germanie. Étude urbanistique.* Publications d'histoire de l'art et d'archéologie de l'Université catholique de Louvain 103. Louvain-la-Neuve: UCL, Presses universitaires de Louvain.

Guillier, Gérard, Axel Levillayer, Bénédicte Pradat, Valérie Deloze, and Philippe Forré. 2015. 'Des poteaux, des greniers et des graines. Une zone de stockage de masse à La Tène C2/D1a au "Clos des Primevères" à Entrammes (Mayenne)'. *Revue archéologique de l'Ouest* 32: 177–260. doi:10.4000/rao.3012.

Haalebos, Jan Kees. 1996. 'Ein Römisches Getreideschiff in Woerden (NL)'. *Jahrbuch Des Römisch-Germanischen Zentralmuseums Mainz* 43: 475–509.

Habermehl, Diederick. 2013. *Settling in a Changing World. Villa Development in the Northern Provinces of the Roman Empire.* Amsterdam Archaeological Studies 19. Amsterdam: Amsterdam University Press. <https://www.jstor.org/stable/j.ctt6wp6vw>.

Habermehl, Diederick, and Jan van Renswoude, eds. 2017. *Duizenden jaren wonen op een rivierduin. Archeologisch onderzoek naar sporen van nederzettingsactiviteit tussen het Mesolithicum en de Nieuwe Tijd te Cuijk-De Nielt.* 2 vols. Nederlandse Archeologische Rapporten 44. Amsterdam / Amersfoort: VUhbs-Vrije Universiteit / Rijksdienst voor het Cultureel Erfgoed. <https://www.cultureelerfgoed.nl/publicaties/duizenden-jaren-wonen-op-een-rivierduin-deel-1–2>.

Halstead, Paul. 1987. 'Traditional and Ancient Rural Economy in Mediterranean Europe: Plus Ça Change?' *The Journal of Hellenic Studies* 107: 77–87. doi:10.2307/630071.

Halstead, Paul. 2014. *Two oxen ahead. Pre-mechanized farming in the Mediterranean.* Chichester: Wiley Blackwell.

Hartmann, Hans-Heinz, and Franz Josef Meyer. 2001. 'Ein *Horreum* in der *Villa Rustica* in Bad Rappenau-Babstadt, Kreis Heilbronn'. *Archäologische Ausgrabungen in Baden-Württemberg* 2001: 127–30.

Hartmann, Hans-Heinz, and Daniel Reichle. 2002. 'Die *Villa Rustica* in Bad Rappenau-Babstadt, Kreis Heilbronn'. *Archäologische Ausgrabungen in Baden-Württemberg* 2002: 135–38.

Haverfield, Francis John. 1924. *The Roman occupation of Britain*. Edited by George MacDonald. Oxford: Clarendon press.

Haverfield, Francis John, and Robin George Collingwood. 1920. 'The Provisionning of Roman Forts'. *Transactions of the Cumberland and Westmorland Antiquarian and Archaeological Society* 20: 127–42. <http://archaeologydataservice.ac.uk/catalogue/adsdata/arch-2055-1/dissemination/pdf/Article_Level_Pdf/tcwaas/002/1920/vol20/tcwaas_002_1920_vol20_0014.pdf>.

Heeren, Stijn. 2009. *Romanisering van rurale gemeenschappen in de* civitas Batavorum. *De casus Tiel-Passewaaij*. Nederlandse Archeologische Rapporten 36. Amersfoort / Amsterdam: Rijksdienst voor het Cultureel Erfgoed / Vrije Universiteit Amsterdam. <http://hdl.handle.net/1871/15248>.

Heeren, Stijn. 2015. 'The depopulation of the Lower Rhine region in 3rd century : an archaeological perspective'. In *The Roman villa of Hoogeloon and the archaeology of the periphery*, edited by Nico Roymans, Ton Derks, and Henk Hiddink, 271–94. Amsterdam archaeological studies 22. Amsterdam: Amsterdam University press.

Heeringen, Robert M. van, Roel Lauwerier, and Henk van der Velde. 1998. *Sporen uit de IJzertijd en de Romeinse tijd in de Hoeksche Waard. Een aanvullend archeologisch onderzoek te Westmaas-Maaszicht, gem. Binnenmaas*. Rapportage Archeologische Monumentenzorg 56. Amersfoort: ROB.

Heimberg, Ursula. 2002–2003. 'Römische Villen an Rhein und Maas'. *Bonner Jahrbücher* 202/203: 57–148. doi:10.11588/bjb.2002.0.42250.

Helly, Benoît. 2014. 'Évolution de la topographie de Vienne, du Ier siècle av. J.-C au VIe siècle ap. J.-C.' In *Vienne*, by Fanny Adjadj, 3:116–48. Carte Archéologique de la Gaule 38. Paris: Académie des Inscriptions et Belles-Lettres.

Hénon, Bénédicte, Ginette Auxiette, Muriel Boulen, Sophie Desenne, Frédéric Gransar, Pascal Le Guen, Claudine Pommepuy, and Bruno Robert. 2002. 'Trois nouveaux sites d'habitat du Hallstatt final/La Tène ancienne dans la vallée de l'Aisne'. In *Découvertes récentes de l'âge du Fer dans le massif des Ardennes et ses marges. Repas des vivants et nourriture pour les morts en Gaule. Actes du XXVe Colloque International de l'AFEAF, Charleville-Mézières, 24–27 mai 2001*, edited by Patrice Méniel and Bernard Lambot, 49–66. Mémoires de La Société Archéologique Champenoise 16. Reims: Société archéologique champenoise.

Herberg, Emil. 1994. 'Römischer Speicher in Bad Rappenau, Landkreis Heilbronn'. *Archäologische Ausgrabungen in Baden-Württemberg* 1994: 196–200.

Herberg, Emil. 1998. 'Weitere Grabungsergebnisse aus dem Bereich des Römischen Gutshofs in Bad Rappenau, Babstadt, Kreis Heilbronn'. *Archäologische Ausgrabungen in Baden-Württemberg* 1998: 158–61.

Herberg, Emil. 1999. 'Die *Villa Rustica* in Bad Rappenau-Babstadt, Kreis Heilbronn. Neue Grabungsergebnisse'. *Archäologische Ausgrabungen in Baden-Württemberg* 1999: 124–28.

Hermansen, Gustav. 1982. *Ostia. Aspects of Roman city life.* Edmonton: University of Alberta Press.

Herz, Peter. 1992. 'Der Aufstand des Iulius Sacrovir (21 n.Chr.). Gedanken zur Römischen Politik in Gallien und ihren Lasten'. *Laverna* 3: 42–93.

Herzig, Hans 1946. 'Die Rekonstruktion von Zeughaus und Getreidemagazin in *Vindonissa*'. *Jahresbericht der Gesellschaft Pro Vindonissa* 1945–1946: 40–47. doi: 10.5169/seals-267544.

Heuzé, Gustave. 1876. *Nouveau manuel complet des constructions agricoles traitant des matériaux et de leur emploi dans les constructions destinées au logement des cultivateurs, des animaux et des produits agricoles dans les petites, moyennes et les grandes exploitations.* Manuels Roret. Paris: Librairie encyclopédique de Roret. <http://gallica.bnf.fr/ark:/12148/bpt6k6534863r/>.

Hinz, Hermann. 1970. 'Zur Bauweise der Villa rustica'. In *Germania Romana III. Römisches Leben auf germanischem Boden*, edited by Hermann Hinz, 15–25. Gymnasium 7. Heidelberg: Carl Winter Verlag.

Hodder, Ian. 1982. *The Iron Age and Romano-British Settlement at Wendens Ambo, Essex.* Passmore Edwards Museum.

Hopkins, Keith. 1980. 'Taxes and Trade in the Roman Empire (200 B.C.–A.D. 400)'. *The Journal of Roman Studies* 70: 101–25. <https://doi.org/10.2307/299558>.

Huitorel, Guillaume. 2017. 'Stocker les céréales dans les établissements ruraux du nord de la Gaule à l'époque romaine. Essai d'identification des modes de stockage entre le Ier et le IVe siècle ap. J.-C.' In *Les céréales dans le monde antique. Regards croisés sur les stratégies de gestion des cultures, de leur stockage et de leur mode de consommation*, edited by Adeline Bats, 217–38. NeHeT. Revue numérique d'égyptologie 5. Paris / Bruxelles: Université Paris-Sorbonne / Université Libre de Bruxelles. Online: <http://sfe-egyptologie.website/images/NEHET5/13-NEHET%205-HUITOREL.pdf>.

Hüssen, Claus-Michael. 2000. *Die römische Besiedlung im Umland von Heilbronn.* Forschungen und Berichte zur Vor- und Frühgeschichte in Baden-Württemberg 78. Stuttgart: K. Theiss.

Jaccottey, Luc, Natàlia Alonso, Sylvie Deffressigne, Caroline Hamon, Stéphanie Lepareux-Couturier, Vérane Brisotto, Sophie Galland-Crety, et al. 2013. 'Le passage des meules va-et-vient aux meules rotatives en France'. In *L'âge du Fer en Europe. Mélanges offerts à Olivier Buchsenschutz*, edited by Sophie Krausz, Anne Colin, Katherine Gruel, Ian B. Ralston, and Thierry Dechezleprêtre, 405–15. Mémoires – Ausonius 32. Bordeaux: Ausonius.

Jakab, Eva. 2014. '*Horrea*, sûretés et commerce maritime dans les archives des *Sulpicii*'. In Inter cives necnon peregrinos. *Essays in honour of Boudewijn Sirks*, edited by Johannes J. Hallebeek, Martin Schermaier, Roberto Fiori, Ernest Metzger, and Jean-Pierre Coriat, 331–50. Göttingen: V & R unipress.

Jahier, Ivan, ed. 2011. *L'enceinte des premier et second âges du Fer de La Fosse Touzé, Courseulles-sur-Mer, Calvados. Entre résidence aristocratique et place de collecte monumentale*. Documents d'archéologie française 104. Paris: Éditions de la Maison des sciences de l'homme.

Jansen, Richard, and Stijn van As. 2012. 'Structuring the Landscape in Iron Age and Roman Period (500 BC–AD 250): The Multi-Period Site Oss-Horzak'. *Analecta Praehistorica Leidensia* 43/44: 95–111. <https://openaccess.leidenuniv.nl/bitstream/handle/1887/32927/APL_43-44_08_Jansen-Van%20As.pdf>.

Jansen, Richard, and Harry Fokkens. 2010. 'Central Places in the 1st and 2nd Century AD in the Maaskant-Region (Southern Netherlands). Reinterpreting the Roman Settlement of Oss-Westerveld'. *Siedlungs- und Küstenforschung im südlichen Nordseegebiet* 33: 68–81. <http://hdl.handle.net/1887/17796>.

Jenter, Susanne, and Jobst M. Wippern. 1997. 'Prospektionsergebnisse von einer Römischen Villa Rustica bei Kirchberg'. *Archäologie im Rheinland* 1996: 177–80.

Johnson, Anne. 1983. *Roman forts of the 1st and 2nd centuries AD in Britain and the German provinces*. London: A. and C. Black.

Johnson, Anne. 1987. *Römische Kastelle des 1. und 2. Jahrhunderts n. Chr. in Britannien und in den germanischen Provinzen des Römerreiches*. Edited by Dietwulf Baatz. Translated by Gabriele Schulte-Holtey. Kulturgeschichte der antiken Wel 37. Mainz: Ph. von Zabern.

Jones, Glynis. 1996. 'Distinguishing Food from Fodder in the Archaeobotanical Record'. *Environmental Archaeology* 1 (1): 95–98. doi:10.1179/env.1996.1.1.95.

Joulin, Léon. 1902. 'Les établissements gallo-romains de la plaine de Martres-Tolosanes'. *Mémoires présentés par divers savants à l'Académie des inscriptions et belles-lettres de l'Institut de France. Première série, Sujets divers d'érudition* 11 (1): 219–516. doi:10.3406/mesav.1902.1085.

Kehne, Peter. 2008. 'Zur Strategie und Logistik römischer Vorstösse in die Germania : Die Tiberiusfeldzüge der Jahre 4 und 5 n. Chr.' In *Rom auf dem Weg nach Germanien: Geostrategie, Vormarschtrassen und Logistik. Internationales Kolloquium in Delbrück-Anreppen vom 4. bis 6. November 2004*, edited by Johann-Sebastian Kühlborn. Bodenaltertümer Westfalens 45. Mainz: Ph. von Zabern.

Kloft, Hans. 2016. 'Antike Getreidespeicher – Ein Werkstattbericht'. In *Wirtschaftsbauten in der Antiken Stadt. Internationales Kolloquium 16.-17. November 2012 Karlsruhe*, edited by Ulrich Fellmeth, Jürgen Krüger, Karlfriedrich Ohr, and Jürgen J. Rasch, 33–43. Materialien Zu Bauforschung Und Baugeschichte 20. Karlsruhe: KIT Scientific Publishing. doi.org :10.5445/KSP/1000055788.

Knörzer, Karl-Heinz. 1970. *Novaesium IV. Römerzeitliche Pflanzenfunde aus Neuss*. Limesforschungen 10. Berlin: G. Mann.

Koehler, Sébastien. 2003. 'Vergers antiques dans les campagnes péri-urbaines : le cas de Reims'. In *Cultivateurs, éleveurs et artisans dans les campagnes de Gaule romaine. Actes du VIe Colloque AGER (Compiègne, 5–7 juillet 2002)*, edited by Sébastien Lepetz and Véronique Matterne, 37–46. Revue archéologique de Picardie, 1/2. Amiens: Revue archéologique de Picardie. doi:10.3406/pica.2003.2355.

Kooistra, Laura I. 1996. *Borderland farming. Possibilities and limitations of farming in the Roman period and early Middle Ages between the Rhine and Meuse*. Assen: Van Gorcum.

Kooistra, Laura I. 2009. 'The provenance of Cereals for the Roman Army in the Rhine Delta. Based on Archaeobotanical Evidence'. In *Kelten am Rhein. Akten des dreizehnten Internationalen Keltologiekongresses 23. bis 27. Juli 2007 in Bonn = Proceedings of the Thirteenth International Congress of Celtic Studies*, edited by Stefan Zimmer, 1:219–37. Beihefte der Bonner Jahrbücher 58. Mainz: Ph. von Zabern.

Kooistra, Laura I. 2012. 'Die pflanzlichen Grundnahrungsmittel der Rheinarmee vor und nach der Gründung der Germania inferior'. In *Verzweigungen. Eine Würdigung für A.J. Kalis und J. Meurers-Balke*, edited by Astrid Stobbe and Ursula Tegtmeier, 171–87. Frankfurter archäologische Schriften 18. Bonn: Habelt.

Kooistra, Laura I., and Stijn Heeren. 2007. 'Het verhaal van een afgebrande graanschuur'. In *Een Bataafse gemeenschap in de wereld van het Romeinse rijk. Opgravingen te Tiel-Passewaaij*, edited by Nico Roymans, Ton Derks, and Stijn Heeren, 167–76. Utrecht: Matrijs.

Kossack, Georg, and Hans Baessler. 2001. 'Ein Speicherbau aus römischer Verwandtschaft in Archsum (Sylt)'. In *'... Trans Albim fluvium'. Forschungen zur vorrömischen kaiserzeitlichen und mittelalterlichen Archäologie. Festschrift für Achim Leube zum 65. Geburtstag*, edited by Michael Meyer, 327–35. Internationale Archäologie. Studia honoraria 10. Rahden: Leidorf.

Kossack, Georg, Ole Harck, and Joachim Reichstein. 1974. 'Zehn Jahre Siedlungsforschung in Archsum Auf Sylt'. *Bericht der Römisch-Germanischen Kommission* 55: 261–377.

Kreuz, Angela. 1994. 'Landwirtschaft und ihre ökologischen Grundlagen in den Jahrhunderten um Christi Geburt: Zum Stand der Naturwissenschaftlichen Untersuchungen in Hessen'. *Berichte der Kommission für Archäologische Landesforschung in Hessen* 3: 59–91.

Kron, Geoffrey. 2008. 'The Much Maligned Peasant. Comparative Perspectives on the Productivity of the Small Farmer in Classical Antiquity'. In *People, land, and politics. Demographic developments and the transformation of Roman Italy 300 BCE–CE 14*, edited by Luuk de Ligt and Simon Northwood, 71–119. Mnemosyne. Supplementum 303. Leiden: Brill.

Krünitz, Johann Georg. 1788. 'Korn-Boden'. In *Oekonomische Encyklopädie oder allgemeines System der Staats- Stadt- Haus- und Landwirthschaft*, edited by Johann Georg Krünitz, 44:766–942. Berlin: J. Pauli. <http://www.kruenitz1.uni-trier.de/>.

Lambot, Bernard, and Patrice Méniel. 1993. *Le site protohistorique d'Acy-Romance (Ardennes)*. Mémoires de la Société archéologique champenoise 7 / Dossier de Protohistoire 4. Paris: Éd. du CNRS.

Lamm, Susanne, and Patrick Marko, eds. 2017. 'Dossier: *Horreum* – Internationales Symposium zu römischen Speicherbauten im Alpenraum [Akten des Symposiums "*Horrea*" am Institut für Archäologie der Universität Graz, 1.–2. Oktober 2015]'. *Römisches Österreich* 40: xi-xii;1–100.

Lang, Felix. 2009. 'Ernteerträge nördlich der Alpen in Römischer Zeit – Überlegungen zur Leistungsfähigkeit der Landwirtschaft und zu den Auswirkungen des Butser Ancient Farm Project'. *Archäologisches Korrespondenzblatt* 39 (3): 393–407.

Lange, Adolf Gustaaf. 1990. *Plant remains from a native settlement at the Roman frontier: De Horden near Wijk bij Duurstede. A numerical approach*. Nederlandse Oudheden 13. Amersfoort: ROB. <https://easy.dans.knaw.nl/ui/datasets/id/easy-dataset:51973>.

Lange, Adolf Gustaaf, Liesbeth Theunissen, Jos Deeben, Jan van Doesburg, and Tessa de Groot, eds. 2014. *Huisplattegronden in Nederland. Archeologische Sporen van het Huis*. Amersfoort: Rijksdienst voor het Cultureel Erfgoed.

Le Goff, Elven. 2016a. 'L'agglomération artisanale et commerciale de Kergolvez à Quimper (Finistère)'. In *Les premières villes de l'Ouest. Agglomérations gauloises de Bretagne et Pays de la Loire. Catalogue d'exposition (30 avril 2016 – 31 mars 2017, Musée Archéologique de Jublains)*, edited by Stephan Fichtl, Elven Le Goff, Agathe Mathiaut-Legros, and Yves Menez, 149–59. Jublains: Musée archéologique de Jublains.

Le Goff, Elven. 2016b. 'L'*Oppidum* de Moulay, capitale gauloise des Aulerques Diablintes'. In *Les premières villes de l'Ouest. Agglomérations gauloises de Bretagne et Pays de la Loire. Catalogue d'exposition (30 avril 2016 – 31 mars 2017, Musée Archéologique de Jublains)*, edited by Stephan Fichtl, Elven Le Goff, Agathe Mathiaut-Legros, and Yves Menez, 123–31. Jublains: Musée archéologique de Jublains.

Lehar, Hannes. 2015. 'Hölzerne Hypokaustheizungen an Lippe und Main?' *Germania : Anzeiger der Römisch-Germanischen Kommission des Deutschen Archäologischen Instituts* 93: 259–75. doi:10.11588/ger.2015.34001.

Lepaumier, Hubert, ed. 2012. *Fleury-Sur-Orne, Calvados, Les Mézerettes – ZL 7. Deux habitats enclos et une nécropole du second âge du Fer du réseau de fermes reconnu au sud-est de l'agglomération caennaise*. Rapport de fouille d'archéologie préventive. Caen: SRA Normandie / Inrap.

Lepaumier, Hubert. 2015. 'Les productions agricoles au second âge du Fer sur la Plaine de Caen : quelques pistes d'études'. In *L' homme, ses ressources et son environnement, dans l'Ouest de la France à l'âge du fer: actualités de la recherche. Actes du Séminaire archéologique de l'Ouest, 24 mars 2014, Université de Nantes*, edited by Marie-Yvane Daire, 149–63. Mémoire de Géosciences Rennes. Hors-série 9. Rennes: Éditions de Géosciences-Rennes.

Lepaumier, Hubert, Chris-Cécile Vauterin, Elven Le Goff, and James Villaregut. 2011. 'Un réseau de fermes en périphérie caennaise'. In *L'âge du fer en Basse Normandie. Actes du XXXIIIe Colloque international de l'AFEAF (Caen, 20–24 mai 2009)*, edited by Philippe Barral, Bernard Dedet, Fabien Delrieu, Pierre Giraud, Isabelle Le Goff, Stéphane Marion, and Anne Villard-Le Tiec, 1:139–58. Annales littéraires de l'Université de Besançon 883. Besançon: Presses universitaires de Franche-Comté.

Lerouxel, Francois. 2017. 'L'Italie républicaine est-elle encore au centre de l'histoire économique romaine ? [à propos de J.-Fr. Chemain, *L'économie romaine en Italie à l'époque républicaine*. – Paris : Picard, 2016]'. *Revue des Études Anciennes* 119 (1): 197–207.

Leveau, Philippe. 2003. 'Inégalités régionales et développement économique dans l'Occident romain (Gaules, Afrique et Hispanie)'. In *Itinéraire de Saintes à Dougga. Mélanges offerts à Louis Maurin*, edited by Jean-Pierre Bost, Jean-Michel Roddaz, and Francis Tassaux, 327–53. Mémoires – Ausonius 9. Bordeaux: Ausonius.

Leveau, Philippe. 2007. 'The western provinces'. In *The Cambridge economic history of the Greco-Roman world*, edited by Walter Scheidel, Ian Morris, and Richard P. Saller, 651–70. Cambridge: Cambridge University Press. doi:10.1017/CHOL9780521780537.025

Levillayer, Axel, Aurélia Borvon, and Jean-Yves Hunot. 2013. 'Du Hallstatt final à La Tène ancienne en Anjou : les batteries de structures de stockage de Luigné et de Cizay-la-Madeleine (Maine-et-Loire)'. *Revue archéologique de l'Ouest* 30: 113–42. doi:10.4000/rao.2086.

Lodwick, Lisa. 2017. 'Arable Farming, Plant Foods and Resources'. In *New Visions of the Countryside of Roman Britain. Volume 2. The Rural Economy of Roman Britain*, by Martyn Allen, Lisa Lodwick, Tom Brindle, Michael Gordon Fulford, and Alex Smith, 11–84. Britannia Monograph Series 30. London: Society for the Promotion of Roman Studies.

Londen, Heleen van. 2006. 'Midden-Delfland. The Roman Native Landscape Past and Present'. Dissertation, Amsterdam: Universiteit van Amsterdam. <https://hdl.handle.net/11245/1.260370>.

Luksen-IJtsma, Annemarie. 2009. *Oudenrijnseweg. Archeologisch onderzoek van een inheems-Romeinse nederzetting uit de eerste eeuw na Chr. en een vlasrootcomplex uit de twaalfde eeuw na Chr.in De Meern, gemeente Utrecht*. Basisrapportage Archeologie 25. Utrecht: Team Erfgoed Gemeente Utrecht. <https://doi.org/10.17026/dans-zn4-qwmm>.

Lüning, Jens, and Jutta Meurers-Balke. 1980. 'Experimenteller Getreideanbau im Hambacher Forst, Gemeinde Elsdorf, Kr. Bergheim/Rheinland'. *Bonner Jahrbücher* 180: 305–44.

Maguer, Patrick, and Dorothée Lusson. 2009. 'Fermes, hameaux et résidences aristocratiques entre Loire et Dordogne'. In *Les Gaulois entre Loire et Dordogne. Actes du XXXIe colloque international de l'AFEAF, 17–20 mai 2007, Chauvigny (Vienne, F)*, edited by Isabelle Bertrand, Alain Duval, and José Gomez de Soto, 423–59. Mémoire – Société de recherches archéologiques de Chauvigny 34. Chauvigny: Association des publications chauvinoises.

Malrain, François. 2000. 'Fonctionnement et hiérarchies des fermes dans la société gauloise du IIIe siècle à la période romaine. L'apport des sites de la moyenne vallée de l'Oise'. Thèse de doctorat, Paris: Université Paris 1 Panthéon-Sorbonne.

Malrain, François, Geertrui Blancquaert, and Thierry Lorho, eds. 2013. *L'habitat rural du second âge du Fer. Rythmes de création et d'abandon au nord de la Loire*. Recherches archéologiques 7. Paris: INRAP / éd. du CNRS .

Malrain, François, Frédéric Gransar, Véronique Matterne, and Isabelle Le Goff. 1996. 'Une ferme de La Tène D1 et sa nécropole : Jaux "Le Camp du Roi" (Oise)'. *Revue archéologique de Picardie* 3–4: 245–306. doi:10.3406/pica.1996.2189.

Malrain, François, Denis Maréchal, Marjolaine De Muylder, Sébastien Lepetz, Patrice Méniel, and Véronique Zech-Matterne. 2017. 'La Vallée de l'Oise'. In Gallia Rustica. *Les campagnes du nord-est de la Gaule, de la fin de l'âge du Fer à l'Antiquité tardive. Tome 1. Études régionales*, edited by Michel Reddé, 303–52. Mémoires – Ausonius 49. Bordeaux: Ausonius.

Malrain, François, Estelle Pinard, Stéphane Gaudefroy, Nathalie Buchez, Cyrille Chaidron, and Stéphane Dubois. 2016. 'Un établissement agricole et ses sépultures du IIIe siècle av. n. è. au IIe s. de n. è. à Poulainville (Somme)'. *Revue Archéologique de Picardie*, 3–4: 5–232.

Manning, William H. 1975. 'Roman Military Timber Granaries in Britain'. *Saalburg-Jahrbuch* 32: 105–29.

Marin, Brigitte, and Catherine Virlouvet, eds. 2004. *Nourrir les cités de Méditerranée. Antiquité-Temps modernes*. L'atelier méditerranéen. Paris / Aix-en-Provence / Madrid: Maisonneuve et Larose / MMSH / Universidad National de Educación a Distancia.

Marin, Brigitte, and Catherine Virlouvet, eds. 2008a. 'Entrepôts et trafics annonaires en Méditerranée'. *Mélanges de l'École Française de Rome – Antiquité* 120 (1): 109–38.

Marin, Brigitte, and Catherine Virlouvet, eds. 2008b. 'Entrepôts et trafics annonaires en Méditerranée'. *Mélanges de l'École Française de Rome. Italie et Méditerranée* 120 (2): 473–606.

Marin, Brigitte, and Catherine Virlouvet, eds. 2008c. 'Entrepôts et trafics annonaires en Méditerranée'. *Mélanges de l'École Française de Rome. Moyen Âge* 120 (1): 5–72.

Marin, Brigitte, and Catherine Virlouvet, eds. 2008d. 'Entrepôts et trafics annonaires en Méditerranée (2)'. *Mélanges de l'École Française de Rome – Antiquité* 120 (2): 275–336.

Marin, Brigitte, and Catherine Virlouvet, eds. 2016. *Entrepôts et trafics annonaires et Méditerranée. Antiquité – Temps modernes*. Collection de l'École française de Rome 522. Rome: École française de Rome.

Marinval, Philippe. 1993. 'Deux amas de grains de blé carbonisés du Haut-Empire romain provenant de Pithiviers-Le-Vieil (Loiret) et de Liniez (Indre)'. *Revue Archéologique du Loiret* 18: 105–19.

Marion, Stéphane. 2013. 'L'économie du IIIe siècle a.C., 20 ans après'. In *L'âge du Fer en Europe. Mélanges offerts à Olivier Buchsenschutz*, edited by Sophie Krausz, Anne Colin, Katherine Gruel, Ian B. Ralston, and Thierry Dechezleprêtre, 361–69. Mémoires – Ausonius 32. Bordeaux: Ausonius.

Martinat, Monica. 2008. 'L'organisation des entrepôts de l'Abondance à Lyon au XVIIe siècle'. *Mélanges de l'École française de Rome – Italie et Méditerranée* 120 (2): 581–93. doi:10.3406/mefr.2008.10566.

Mathieu de Dombasle, Christophe-Joseph-Alexandre. 1862. *Traité d'agriculture, publié sur le manuscrit de l'auteur par Ch. de Meixmoron de Dombasle, son petit-fils. Deuxième partie. Pratique agricole II*. Paris: Vve Bouchard-Huzard / Libraire agricole.

Mathiot, Dimitri. 2012. 'Les habitats ruraux des cinq derniers siècles avant notre ère entre le cours de la Somme et le delta Meuse-Rhin : formes et organisations'. *Revue du Nord* 94 (398): 7–26. doi:10.3917/rdn.398.0007.

Matterne, Véronique. 1997. 'Étude d'une concentration de semences carbonisées provenant d'une structure du Xe siècle, place des Hallettes à Compiègne (Oise)'. In *Fouilles de sauvetage sous la place du marché à Compiègne (Oise) -1991/1993. L'évolution urbaine de l'aire palatiale du haut Moyen-Âge aux marchés médiéval et moderne*, edited by Martin Petitjean, 219–28. Revue archéologique de Picardie. Numéro spécial 13. Amiens: Revue archéologique de Picardie. doi:10.3406/pica.1997.1949.

Matterne, Véronique. 2000. 'Évolution des productions agricoles durant l'âge du Fer dans le nord de la France'. In *Les installations agricoles de l'âge du Fer en France septentrionale*, edited by Gertrude Blancquaert and Stéphane Marion, 129–46. Études d'histoire et archéologie 6. Paris: Éd. Rue d'Ulm, Presses de l'École normale supérieure.

Matterne, Véronique. 2001. *Agriculture et alimentation végétale durant l'âge du fer et l'époque gallo-romaine en France septentrionale*. Archéologie des plantes et des animaux 1. Montagnac: Éd. M. Mergoil.

Matterne, Véronique, Jean-Hervé Yvinec, and Dominique Gemehl. 1998. 'Stockage de plantes alimentaires et infestation par les insectes dans un grenier incendié de la fin du IIe siècle après J.-C. à Amiens (Somme)'. *Revue Archéologique de Picardie* 3 (1): 93–122. doi:10.3406/pica.1998.2290.

Mauné, Stéphane, and Jean-Marie Paillet. 2003. 'Stockage et transformation des céréales dans l'économie rurale de Gaule narbonnaise (Ier–IIe siècle apr. J.-C.). L'exemple des moulins hydrauliques de Vareilles et de l'Auribelle-Basse (Hérault)'. In *Le traitement des récoltes. Un regard sur la diversité du néolithique au présent*, edited by Patricia C. Anderson, Linda S. Cummings, Thomas K. Schippers, and Bernard Simonel, 295–326. Rencontres internationales d'archéologie et d'histoire d'Antibes 23. Antibes: Éd. APDCA.

Menez, Yves. 2009. 'Le Camp de Saint-Symphorien à Paule (Côtes d'Armor) et les résidences de l'aristocratie du second âge du Fer en France septentrionale'. Thèse de doctorat, Paris: Université Paris 1 Panthéon-Sorbonne.

Merten, Jürgen, ed. 2005. *Rettet das archäologische Erbe in Trier. Zweite Denkschrift der Archäologischen Trier-Kommission.* Schriftenreihe des Rheinischen Landesmuseums Trier 31. Trier: Rheinisches Landesmuseum.

Metzler, Jeannot, Johny Zimmer, and Lothar Bakker. 1981. *Ausgrabungen in Echternach.* Luxembourg: Publication du Ministère des Affaires Culturelles et de la Ville d'Echternach.

Moreau, Clément, and Fanny Granier. 2012. 'Val-de-Reuil (Eure), "Le Chemin aux Errants". Une occupation rurale du premier et du second âge du Fer à la confluence de l'Eure et de la Seine'. In *Journées archéologiques de Haute-Normandie. Évreux, 6–8 mai 2011*, edited by Marie-Clotilde Lequoy and Luc Liogier, 27–48. Mont-Saint-Aignan: Publications des Universités de Rouen et du Havre.

Moreau, Clément, Fanny Granier, and Emilie Dubreucq. 2015. 'Stockage et transport fluvial des céréales au Second âge du Fer : l'exemple de Val-de-Reuil (Eure)'. In *Les Gaulois au fil de l'eau. Actes du 37e colloque international de l'AFEAF (Montpellier, 8–11 mai 2013). Volume 1, Communications*, edited by Fabienne Olmer and Réjane Roure, 539–68. Mémoires – Ausonius 39. Bordeaux: Ausonius, 2015.

Morris, Pat. 1979. *Agricultural buildings in Roman Britain*. BAR. British series 70. Oxford: British archaeological reports.

Nicolier, Anelise. 2015. 'Le grenier d'abondance de Lyon, 6 quai Saint-Vincent (1722–1777)'. Online: <https://halshs.archives-ouvertes.fr/halshs-01251873/>.

Nouvel, Pierre. 2004. 'Des terroirs et des hommes. Dynamique des organisations spatiales dans le Bassin de l'Yonne moyenne et leur évolution de la fin de l'Âge du Bronze au haut Moyen Âge'. Thèse de doctorat, Dijon: Université de Bourgogne.

Nouvel, Pierre. 2016. 'Entre Ville et Campagne. Formes de l'occupation et élites gallo-romaines dans le Centre-Est de la Gaule. Apport de 20 années de prospections et de fouilles archéologiques'. Habilitation à diriger des recherches, Besançon: Université de Franche-Comté.

Othenin-Girard, Blaise. 1999. 'Alle JU, Les Aiges'. *Jahrbuch der schweizerischen Gesellschaft für Ur- und Frühgeschichte* 82: 277.

Othenin-Girard, Blaise. 2002. 'Alle JU, Les Aiges'. *Jahrbuch der schweizerischen Gesellschaft für Ur- und Frühgeschichte* 85: 306.

Othenin-Girard, Blaise, and Mustapha Elyaqtine. 2002. *Greniers gallo-romains et vestiges funéraires du Bronze final à Alle, Les Aiges (Jura, Suisse) : Fouilles 2001*. Archéologie et Transjurane 95. Porrentruy: Office du patrimoine historique.

Ouzoulias, Pierre. 2006. 'L'économie agraire de la Gaule : aperçus historiographiques et perspectives archéologiques'. Thèse de doctorat, Besançon: Université de Franche-Comté. <https://tel.archives-ouvertes.fr/tel-00011567/>.

Oyen, Astrid Van. 2015. 'The Moral Architecture of Villa Storage in Italy in the 1st c. B.C.' *Journal of Roman Archaeology* 28 (1): 97–123. doi:10.1017/S1047759415002421.

Päffgen, Bernd. 2000. 'Villa rustica und Burgus auf dem Steinacker in Jülich-Kirchberg'. In *Millionen Jahre Geschichte, Fundort Nordrhein-Westfalen. Begleitbuch zur Landesausstellung Fundort Nordrhein-Westfalen, Millionen Jahre Geschichte, Köln, Römisch-Germanisches Museum, 17. März bis 27. August 2000, Münster, Westfälisches Museum für Archäologie, 12. November 2000 bis 11. März 2001, Nijmegen, Museum Het Valkhof, Mai bis August 2001*, edited by Heinz Günter Horn, 283–84. Schriften zur Bodendenkmalpflege in Nordrhein-Westfale. Mainz: Ph. von Zabern.

Pals, Jan Peter, Vera Beemster, and Aart Noordam. 1989. 'Plant Remains from the Roman Castellum Praetorium Agrippinae Near Valkenburg (prov. of Zuid-Holland)'. In *Archäobotanik: Symposium der Universität Hohenheim (Stuttgart) vom 11.-16. Juli 1988*, edited by Udelgard Körber-Grohne and Hansjörg Küster, 117–33. Dissertationes botanicae 133. Berlin: J. Cramer.

Pals, Jan Peter, and Tom Hakbijl. 1992. 'Weed and Insect Infestation of a Grain Cargo in a Ship at the Roman Fort of Laurium in Woerden (Province of Zuid-Holland)'. *Review of Palaeobotany and Palynology* 73 (1–4): 287–300. doi:10.1016/0034-6667(92)90064-N.

Papi, Emanuele, and Francesco Martorella. 2007. 'Il grano della *Tingitana*'. In *Supplying Rome and the Empire. Proceedings of an international seminar held at Siena-Certosa di Pontignano on May 2–4, 2004, on Rome, the provinces, production and distribution*, edited by Emanuele Papi, 85–96. Journal of Roman archaeology. Supplementary series 69. Portsmouth: Journal of Roman archaeology.

Parmentier, Antoine Augustin. 1789. *Mémoire sur les avantages que le royaume peut retirer de ses grains, considérés sous leurs différens rapports avec l'agriculture, le commerce, la meunerie et la boulangerie avec le mémoire sur la nouvelle manière de construire les moulins à farine, qui a remporté le prix de l'Académie royale des sciences en 1785, par M. Dransy, ingénieur du Roi,... on y a joint un manuel sur la manière de traiter les grains, et d'en faire du pain.* Paris: Chez Barrois l'aîné.

Pavlovic, Andelko, ed. 2011. *Archeologisch onderzoek aan de Uithofslaan, gemeente Den Haag. Deel 1: Sporen van bewoning uit de ijzertijd (vindplaats 6) en de Romeinse tijd* (*vindplaats 3*). Haagse Archeologische Rapportage 1122. Den Haag: Gemeente Den Haag, Dienst Stadsbeheer, Afdeling Archeologie. doi:10.17026/dans-2cm-nwpm.

Pellegrino, Vincenzo, Stéphane Mauné, and Véronique Mathieu. 2017. 'Les bâtiments de type "tour" en contexte rural dans les Gaules. Bâtiments d'exploitation ou d'agrément ?' In *Produire, transformer et stocker dans les campagnes des Gaules romaines. Problèmes d'interprétation fonctionnelle et économique des bâtiments d'exploitation et des structures de production agro-pastorale. Colloque international AGER XI (Clermont-Ferrand, 11–13 juin 2014)*, edited by Frédéric Trément, 181–211. Aquitania. Supplément. Bordeaux: Fédération Aquitania.

Pellegrino, Vincenzo. 2017. 'Granai e spazi per lo stoccaggio e per il trattamento dei cereali nelle villae rusticae vesuviane'. *Mélanges de l'École française de Rome – Antiquité* 129 (2): online. doi:10.4000/mefra. 4506>.

Petculescu, Liviu. 1987. 'Roman Military Granaries in Dacia'. *Saalburg-Jahrbuch* 43: 66–76.

Petit, Jean-Paul. 2005. *Bliesbruck-Reinheim: Celtes et Gallo-Romains en Moselle et en Sarre*. Hauts lieux de l'histoire. Paris: Éd. Errance.

Petrikovits, Harald von. 1975. *Die Innenbauten römischer Legionslager während der Prinzipatszeit*. Abhandlungen der Rheinisch-Westfälischen Akademie der Wissenschaften 56. Opladen: Westdeutscher Verlag.

Philippe, M., ed. 1993. *La déviation de la RN 12 Houdan-Bazainville (Yvelines), rapport de fouilles, S.R.A.I.F., Yvelines, Paris/Versailles*. Rapport de fouilles. Paris / Versailles: SRA Île-de-France / DDE Yvelines.

Philippe, M., L. Ajot, M.-F. André, G. Boulay, P. Brunet, V. Gonzalez, N. Paccard, and C. Piozzoli. 1994. 'Opération intercommunale: Déviation RN 12, Houdan-Bazainville'. In *Bilan Scientifique de La DRAC Île de France*, 101–3. Houdan: SRA Île-de-France.

Piening, Ulrike. 1988. 'Verkohlte Pflanzenreste aus zwei römischen Gutshöfen bei Bad Dürkheim (Pfalz)'. In *Der Prähistorische Mensch und seine Umwelt. Festschrift für Udelgard Körber-Grohne zum 65. Geburtstag*, edited by Hansjörg Küster, 325–40. Forschungen und Berichte zur Archäologie in Baden-Württemberg 30. Stuttgart: Kommissionsverlag K. Theiss.

Pilon, Fabien. 1998. 'La fabrication de monnaies d'imitation frappées et coulées à Châteaubleau (Seine-et-Marne) au IIIe Siècle après J.-C.' *Trésors Monétaires* 17: 77–106.

Pilon, Fabien. 2003. 'La villa gallo-romaine du "Bois du Châtel" (Vieux-Champagne, Seine-et-Marne) : production de fausses monnaies en milieu rural'. *Revue Archéologique de Picardie* 1–2: 177–83. doi:10.3406/pica.2003.2364.

Pion, Patrick, Claudine Pommepuy, Ginette Auxiette, Bénédicte Hénon, and Frédéric Gransar. 1997. 'L'oppidum de Condé-sur-Suippe/Variscourt (Aisne) (fin IIe – début Ier siècle av. J.-C.) : approche préliminaire de l'organisation fonctionnelle d'un quartier artisanal'. In *Espaces physiques, espaces sociaux dans l'analyse interne des sites du Néolithique à l'âge du Fer. Actes du 119e Congrès National des Sociétés Historiques et Scientifiques, Amiens, 26–30 octobre 1994*, edited by Ginette Auxiette, Lamys Hachem, Bruno Robert, and Bocquet, 275–309. Congrès National des Sociétés Historiques et Scientifiques. Actes 119. Paris: Éd. du CTHS.

Poirier, Maxence. 2011. 'Le bâtiment A2 de l'état 5. Argumentaire contradictoire des différentes restitutions'. In Durocortorum*, rue Maucroix. Un quartier excentré d'une capitale de province romaine (fin du 1er s. av. J.-C.-début du 4e s.)*, by Philippe Rollet, Francois Berthelot, Guillaume Florent, and Emilie Jouhet, 130–33. Bulletin de la Société archéologique champenoise, 104.4. Reims: Société archéologique champenoise.

Polak, Marinus, and Laura I. Kooistra. 2013. 'A Sustainable Frontier? The Establishment of the Roman Frontier in the Rhine Delta. Part 1: From the End of the Iron Age to the Death of Tiberius (c. 50 BC–AD 37)'. *Jahrbuch des Römisch-Germanischen Zentralmuseums Mainz* 60 (2): 355–458. doi:10.11588/jrgzm.2013.2.20315.

Postharvest Food Losses in Developing Countries. 1978. Washington (D.C.): National Academy of Sciences. <http:// archive.org/details/Post_Harvest_Food_Losses_in_Developing_Countries>.

Poux, Matthieu. 2017. 'Le *granarium* des Buissières à Panossas : contribution à l'étude des réseaux d'entrepôts de grande capacité dans les Gaules et les Germanies (Ier–IIIe s. ap. J.-C.),'. In *Produire, transformer et stocker dans les campagnes des Gaules romaines. Problèmes d'interprétation fonctionnelle et économique des bâtiments d'exploitation et des structures de production agro-pastorale. Colloque international AGER XI (Clermont-Ferrand, 11–13 juin 2014)*, edited by Frédéric Trément, 407–34. Aquitania. Supplément 38. Bordeaux: Fédération Aquitania.

Poux, Matthieu, and Aldo Borlenghi. 2016. 'La station de *Bergusium* et le site des Buissières à Panossas (Isère). De la toponymie à l'archéologie'. *Gallia* 73 (1): 133–63.

Purcell, Nicholas. 1995. 'The Roman villa and the landscape of production'. In *Urban society in Roman Italy*, edited by Tim J. Cornell and Kathryn Lomas, 157–84. London: UCL press.

Rabold, Britta. 2001. 'Römisches Landgut oder Kaiserliche Domäne bei Walldorf, Rhein-Neckar-Kreis'. *Archäologische Ausgrabungen in Baden-Württemberg*, 138–42.

Rabold, Britta. 2005. 'Walldorf (HD). Römisches Landgut oder kaiserliche Dömane?' In *Die Römer in Baden-Württemberg. Römerstätten und Museen von Aalen bis Zwiefalten*, edited by Dieter Planck, 356–58. Stuttgart: Theiss.

Raepsaet, Georges, and Marie-Thérèse Raepsaet-Charlier. 2013. 'La Zélande à l'époque romaine et la question des Frisiavons'. *Revue du Nord* 95 (403):209–42. doi:10.3917/rdn.403.0209.

Raepsaet-Charlier, Marie-Thérèse. 1999. 'Les institutions municipales dans les Germanies sous le Haut-Empire : bilan et questions'. In *Cités, municipes, colonies. Les processus de municipalisation en Gaule et en Germanie sous le Haut Empire romain,* edited by Monique Dondin-Payre and Marie-Thérèse Raepsaet-Charlier, 271–352. Publications de la Sorbonne. Histoire ancienne et médiévale 53. Paris: Publications de la Sorbonne.

Reddé, Michel, ed. 2009. *Oedenburg. Fouilles françaises, allemandes et suisses à Biesheim et Kunheim, Haut-Rhin, France. Volume 1, Les camps militaires julio-claudiens.* Monographien – Römisch-Germanisches Zentralmuseum, 79,1. Mainz: Verlag des Römisch-Germanischen Zentralmuseums.

Reddé, Michel. 2011a. 'Analyse des bâtiments de la rue Maucroix et propositions de restitutions architecturales'. In Durocortorum*, rue Maucroix. Un quartier excentré d'une capitale de province romaine (fin du 1er s. av. J.-C.-début du 4e s.)*, by Philippe Rollet, Francois Berthelot, Guillaume Florent, and Emilie Jouhet, 124–29. Bulletin de la Société archéologique champenoise, 104.4. Reims: Société archéologique champenoise.

Reddé, Michel. 2011b. '"Ut eo terrore commeatus Gallia aduentantes interciperentur" (Tacite, *Hist.* V, 23). La Gaule intérieure et le ravitaillement de l'armée du Rhin.' *Revue des Études Anciennes* 113: 489–509.

Reddé, Michel, ed. 2018a. Gallia Rustica. *Les campagnes du nord-est de la Gaule, de la fin de l'âge du Fer à l'Antiquité tardive. Tome 2. Synthèse.* Mémoires – Ausonius 50. Bordeaux: Ausonius.

Reddé, Michel. 2018b [in press]. 'Genèse et développement des formes de la villa romaine dans le nord-est de la Gaule'. *Jahrbuch Des Römisch-Germanischen Zentralmuseums Mainz.*

Reddé, Michel, Raymond Brulet, Rudolf Fellmann, Jan Kees Haalebos, and Siegmar von Schnurbein, eds. 2006. *L'architecture de La Gaule romaine. Les fortifications militaires.* Documents d'archéologie Française 100. Paris / Bordeaux.

Reneaume. 1730 [1708]. 'Sur la manière de conserver les grains'. *Histoire de l'Académie royale des sciences, avec les mémoires de mathématique & de physique tirez des registres de cette Académie*, Année 1708:63–86.

Rickman, Geoffrey. 1971. *Roman granaries and store buildings.* Cambridge: Cambridge University Press.

Rickman, Geoffrey. 1980. *The corn supply of ancient Rome.* Oxford: Clarendon Press.

Richmond, Ian A., and James McIntyre. 1938–1939. 'The Agricolan Fort at Fenloch'. *Proceedings of the Society of Antiquaries of Scotland* 73:110–54.

Rodbertus, Johan Karl. 1907. 'Per la storia dell'evoluzione agraria di Roma sotto gli imperatori'. In *Biblioteca di storia economica. vol. 2, Parte seconda*, edited by Vilfredo Pareto, 455–508. Milano, Italie: Societa ed. Libraria.

Roger, Jacques, and Philippe Bet. 2005. 'L'établissement rural gallo-romain de Laschamp à Parsac (Creuse)'. *Aquitania* 21: 151–87.

Rollet, Philippe, Francois Berthelot, Guillaume Florent, and Emilie Jouhet. 2011 [2014]. Durocortorum*, rue Maucroix. Un quartier excentré d'une capitale de province ro-*

maine (fin du 1er *s. av. J.-C.-début du 4*^{e} *s.)*. Bulletin de la Société archéologique champenoise, 104.4. Reims: Société archéologique champenoise.

Rösch, Mannfred. 2009. 'Der Inhalt eines horreums von Bad Rappenau, Kreis Heilbronn'. In *Landesarchäologie. Festschrift für Dieter Planck zum 65. Geburtstag*, edited by Jörg Biel, Jörg Heiligmann, and Dirk Krausse, 379–89. Forschungen und Berichte zur Vor- und Frühgeschichte in Baden-Württemberg. Stuttgart: K. Theiss.

Rossiter, Jeremy John. 1978. *Roman farm buildings in Italy*. BAR. International series 52. Oxford: British Archaeological Reports.

Rothé, Marie-Pierre. 2001. *Le Jura*. Carte archéologique de la Gaule 39. Paris.

Roy, Eddy, ed. 2009. *Stockage lié à des productions meunières et dépôt monétaire osisme découverts à Laniscat (Côtes-d'Armor)*. Rapport de fouille d'archéologie préventive. Rennes: SRA Bretagne / Inrap.

Roymans, Nico. 1996. 'The sword or the plough. Regional dynamics in the romanisation of Belgic Gaul and the Rhineland area'. In *From the sword to the plough. Three studies on the earliest romanisation of northern Gaul*, edited by Nico Roymans, 9–126. Amsterdam archaeological studies 1. Amsterdam: Amsterdam University Press.

Roymans, Nico, and Diederick Habermehl. 2011. 'On the origin and development of axial villas with double courtyards in the Latin West'. In *Villa landscapes in the Roman North. Economy, culture and lifestyles*, edited by Nico Roymans and Ton Derks, 83–105. Amsterdam archaeological studies 17. Amsterdam: Amsterdam University Press. <https://www.jstor.org/stable/j.ctt46mx3k.7>.

Ruas, Marie-Pierre. 2003. 'Des grains aux pratiques : le traitement des céréales au XIVe siècle en montagne Noire'. In *Le traitement des récoltes. Un regard sur la diversité du néolithique au présent*, edited by Patricia C. Anderson, Linda S. Cummings, Thomas K. Schippers, and Bernard Simonel, 173–200. Rencontres internationales d'archéologie et d'histoire d'Antibes 23. Antibes: Éd. APDCA.

Ruas, Marie-Pierre, Laurent Bouby, Vanessa Py, and Jean-Paul Cazes. 2005. 'An 11th Century A.D. Burnt Granary at La Gravette, South-Western France: Preliminary Archaeobotanical Results'. *Vegetation History and Archaeobotany* 14 (4):416–26. doi:10.1007/s00334-004-0057-4.

Rudolf, Hans Ulrich. 1989a. 'Bau, Funktion und Bedeutung des Fruchtkastens in der Klosterzeit (1684–1802)'. In *Der Fruchtkasten des Klosters Weingarten 1688–1988*, by Hans Ulrich Rudolf, Norbert Kruse, and Ursula Kaiser, 15–35. Weingartener Hochschulschriften 7. Bergatreute: W. Eppe.

Rudolf, Hans Ulrich. 1989b. 'Die Speicherkapazität des Fruchtkastens'. In *Der Fruchtkasten des Klosters Weingarten 1688–1988*, by Hans Ulrich Rudolf, Norbert Kruse, and Ursula Kaiser, 51–53. Weingartener Hochschulschriften 7. Bergatreute: W. Eppe.

Rupp, Vera, Martin Schmid, and Nicole Boenke. 1998. *Der römische Gutshof 'Im Brückfeld' in Münzenberg-Gambach, Wetteraukreis: Ausgrabungen und Forschungen der Jahre 1994–1998*. Archäologische Denkmäler in Hessen 145. Wiesbaden: Landesamt für denkmalpflege Hessen.

Sagnier, Henry, ed. 1920. *Omnium agricole. Dictionnaire pratique de l'agriculture moderne*. Paris: Librairie Hachette.

Salido Domínguez, Javier. 2008. 'La investigación sobre los *horrea* de época romana: Balance historiográfico y perspectivas de futuro'. *Cuadernos de Prehistoria y Arqueología* 34: 105–24.

Salido Domínguez, Javier. 2011. Horrea Militaria. *El aprovisionamiento de grano al ejército en el occidente del Imperio Romano*. Anejos de Gladius 14. Madrid: Consejo Superior de Investigaciones Científicas : Polifemo Ediciones.

Salido Domínguez, Javier. 2013. 'El abastecimiento de grano a las ciudades hispanorromanas. Producción, almacenaje y gestión'. *Archivo Español de Arqueología* 86: 131–48. doi:10.3989/aespa.086.013.008.

Salido Domínguez, Javier. 2017. *Arquitectura rural romana: graneros y almacenes en el Occidente del Imperio*. Archéologie et histoire romaine 35. Autun: Éd. M. Mergoil.

Salido Domínguez, Javier, and Macarena Bustamante Álvarez. 2014. Pistrina Hispaniae. *Panaderías, molinerías y el artesanado alimentario en la Hispania romana*. Monographies Instrumentum 47. Montagnac: M. Mergoil.

Saunders, Charles Edward. 1904. 'Emmer and Spelt'. *Bulletin, Dominion Experimental Farms and Stations, 2nd Series* 45: 3–16.

Scherer, Peter. 1969. *Reichsstift und Gotteshaus Weingarten im 18. Jahrhundert: Ein Beitrag zur Wirtschaftsgeschichte der südwestdeutschen Grundherrschaft*. Kommission für Geschichtliche Landeskunde in Baden-Württemberg. – Veröffentlichungen. Reihe B.: Forschungen 57. Stuttgart: W. Kohlhammer.

Schinkel, Kees. 1998. 'Unsettled settlement, occupation remains from the Bronze Age and Iron Age at Oss-Ussen. The 1976–1986 excavations'. In *The Ussen project. The first decade of excavations at Oss*, edited by Harry Fokkens, 5–305. Analecta Praehistorica Leidensia 30. Leiden: Leiden University. <https://hdl.handle.net/1887/33828>.

Schubert, Tobias. 2016. 'Pars Fructuaria'. *Studie zu Nebengebäuden mit Speicherfunktion auf römerzeitlichen Villae im Tagebaugebiet Hambacher Forst. Mit Vergleichen aus den germanischen und britannischen Provinzen*. Hamburg: Tredition. <http://kups.ub.uni-koeln.de/7068/>.

Schucany, Caty. 2006. *Die römische Villa von Biberist-Spitalhof/SO (Grabungen 1982, 1983, 1986–1989). Untersuchungen im Wirtschaftsteil und Überlegungen zum Umland*. 3 vols. Ausgrabungen und Forschungen 4. Remshalden: B.A. Greiner.

Séguier, Jean-Marc. 1995. *Marolles-Sur-Seine (Seine-et-Marne), "Le Grand Canton Sud et Le Chemin de Sens" : un gisement archéologique de l'interfluve Seine-Yonne du Paléolithiques supérieur à l'Antiquité tardive*. Rapport de Fouille d'archéologie Préventive. Paris / Bazoches-lès-Bray: SRA Île-de-France / Centre départemental d'archéologie de la Bassée-Saint-Denis.

Séguier, Jean-Marc, Ginette Auxiette, Benoît Clavel, Olivier Maury, and Sylvie Rimbault. 2006–2007. 'Le début du IVe s. av. J.-C. dans l'espace culturel sénonais : les habitats de Bois d'Echalas à Ville-Saint-Jacques et de Beauchamp à Varennes-sur-Seine (Seine-et-Marne) dans le contexte de l'interfluve Seine-Yonne'. *Revue archéologique du Centre de la France* 45–46, online: <https://journals.openedition.org/racf/611>.

Serres, Olivier de. 1600. *Le Théatre d'agriculture et mesnage des champs*. Paris: Jamet Mettayer. <http://bibliotheque-numerique.hortalia.org/items/show/49>.

Sigaut, François. 1978. *Les réserves de grains à long terme. Techniques de conservation et fonctions sociales dans l'histoire*. Travaux et documents 2. Paris / Lille: Éditions de la Maison des sciences de l'homme / Publications de l'Université de Lille III.

Sigaut, François. 1978. 1981. 'Identification des techniques de conservation et de stockage des grains'. In *Les techniques de conservation des grains à long terme. Leur rôle dans la dynamique des systèmes de cultures et des sociétés 2*, edited by Marceau Gast, François Sigaut, and Ariane Bruneton-Governatori, 156–80. Paris: Éditions du CNRS.

Sigaut, François. 1978. 1988. 'A Method for Identifying Grain Storage Techniques and Its Application for European Agricultural History'. *Tools and Tillage* 6 (1): 3–32.

Sigaut, François. 1978. 2012. 'La conservación de las cosechas, treinta años después'. In *Almacenamiento prehispánico del Norte de México al Altiplano Central*, edited by Séverine Bortot, Dominique Michelet, and Véronique Darras. México / Paris: Centro de Estudios Mexicanos y Centroamericanos, Universidad Autónoma de San Luis de Potosí / Laboratoire Archéologie des Amériques, Université de Paris 1 Panthéon Sorbonne.

Sirks, Adriaan Johan Boudewijn. 1991. *Food for Rome. The legal structure of the transportation and processing of supplies for the imperial distributions in Rome and Constantinople*. Studia Amstelodamensia ad epigraphicam, ius antiquum et papyrologicam pertinentia 31. Amsterdam: J.G. Gieben.

Smith, Alexander. 2016. 'Buildings in the Countryside'. In *New Visions of the Countryside of Roman Britain. Volume 1. The Rural Settlement of Roman Britain*, by Alexander Smith, Martyn Allen, Tom Brindle, Michael Gordon Fulford, and Neil Holbrook, 44–74. Britannia Monograph Series 29. London: Society for the Promotion of Roman Studies.

Smith, Alexander, Martyn Allen, Tom Brindle, Michael Gordon Fulford, and Neil Holbrook. 2016. *New visions of the countryside of Roman Britain. Volume 1. The rural settlement of Roman Britain*. Britannia monograph series 29. London: Society for the Promotion of Roman Studies.

Smith, David, and Harry Kenward. 2011. 'Roman Grain Pests in Britain: Implications for Grain Supply and Agricultural Production'. *Britannia* 42: 243–62. doi:10.1017/S0068113X11000031.

Smith, David, and Harry Kenward. 2012. '"Well, Sextus, What Can We Do with This?" The Disposal and Use of Insect-Infested Grain in Roman Britain'. *Environmental Archaeology* 17 (2): 141–50. doi:10.1179/1461410312Z.00000000012.

Smith, John T. 1963. 'Romano-British Aisled Houses'. *Archaeological Journal* 120: 1–30. doi:10.1080/00665983.1963.10854236.

Smith, David, and Harry Kenward. 1997. *Roman villas. A study in social structure.* London: Routledge.

Sommer, Christopher Sebastian. 2005. 'Römische Häuser: 12 m bis zum First. Die villa rustica von Oberndorf-Bochingen'. In *Imperium Romanum: Roms Provinzen an Neckar, Rhein und Donau. Begleitband zur Ausstellung des Landes Baden-Württemberg im Kunstgebäude Stuttgart, 1. Oktober 2005 bis 8. Januar 2006*, 282–85. Stuttgart: Archäologischen Landesmuseum.

Stallknecht, Gilbert F., K.M. Gilbertson, and J.E. Ranney. 1996. 'Alternative Wheat Cereals as Food Grains: Einkorn, Emmer, Spelt, Kamut, and Triticale'. In *Progress in New Crops*, edited by Jules Janick, 156–70. Alexandria, VA: ASHS Press.

Straker, Vanessa. 1984. 'First and second century carbonised cereal grain from Roman London'. In *Plants and ancient man. Studies in palaeoethnobotany. Proceedings of the Sixth Symposium of the International Work Group for Palaeoethnobotany, Groningen, 30 May-3 June 1983*, edited by Willem van Zeist and Willem Arnold Casparie, 323–29. Rotterdam / Boston: A.A. Balkema.

Strickland, Donald. 1987. 'The Granary – Grain capacity'. In *The Roman villa at Lullingstone, Kent. Vol. II*, edited by Geoffroy Wells Meates, 319. Monograph series of the Kent Archaeological Society 3. Maidstone: Kent Archaeological Society.

Tchernia, André. 2011. *Les Romains et le commerce.* Études VIII. Naples: Centre Jean Bérard.

Tchernia, André. 2016. *The Romans and trade.* Translated by James Grieve and Elizabeth Minchin. Oxford Studies on the Roman Economy. Oxford: Oxford University Press, 2016.

Tengström, Emin. 1974. *Bread for the people. Studies of the corn-supply of Rome during the Late Empire.* Skrifter utgivna av Svenska institutet i Rom. Series altera in 8° 12. Stockholm: Stockholms universiteit.

Tessier, Alexandre-Henri. 1793. 'Art. "Conservation Des Grains"'. In *Encyclopédie Méthodique. Agriculture. T. 3* [*Chable-Cytise*], by Alexandre-Henri Tessier, André Thouin, Jean-François Reynier, and Louis-Augustin-Guillaume Bosc, 454–752. Paris: Chez Panckoucke.

Trément, Frédéric, ed. 2017. *Produire, transformer et stocker dans les campagnes des Gaules romaines. Problèmes d'interprétation fonctionnelle et économique des bâtiments d'exploitation et des structures de production agro-pastorale. Colloque international AGER XI (Clermont-Ferrand, 11–13 juin 2014)*. Aquitania. Supplément 38. Bordeaux: Fédération Aquitania.

Van Ossel, Paul. 1992. *Établissements ruraux de l'Antiquité tardive dans le nord de la Gaule*. Gallia. Supplément 51. Paris: Éd. du CNRS.

Van Ossel, Paul, and Ann Defgnée. 2001. *Champion, Hamois. Une villa romaine chez les Condruses. Archéologie, environnement et économie d'une exploitation agricole antique de la Moyenne Belgique*. Études et documents. Archéologie 7. Namur: Ministère de la région wallonne, Direction générale de l'aménagement du territoire, du logement et du patrimoine, Division du patrimoine.

Veen, Marijke van der. 1988. 'Carbonised grain from a Roman granary at South Shields, North East England'. In *Der Prähistorische Mensch und seine Umwelt. Festschrift für Udelgard Körber-Grohne zum 65. Geburtstag*, edited by Hansjörg Küster, 353–65. Forschungen und Berichte zur Archäologie in Baden-Württemberg 30. Stuttgart: Kommissionsverlag K. Theiss.

Veldman, Anouk and Edwin Blom. 2010. *Onder de zoden van Zaltbommel. Een rurale nederzetting en een grafveld uit de Romeinse tijd in het plangebied De Wildeman*. ADC Monografie 8. Amersfoort: ADC ArcheoProjecten.

Vigil-Escalera, Alfonso, Giovanna Bianchi, and Juan Antonio Quirós Castillo, eds. 2013. *Horrea, barns and silos. Storage and incomes in Early Medieval Europe*. Documentos de arqueología medieval 5. Bilbao: Universidad del País Vasco, Servicio editorial = Euskal Herriko Unibertsitatea, Argitalpen Zerbitzua.

Virlouvet, Catherine. 2000. 'Les denrées alimentaires dans les archives des *Sulpicii* de Pouzzoles'. *Cahiers du Centre Gustave Glotz* 11:131–49. doi:10.3406/ccgg.2000.1526.

Virlouvet, Catherine, ed. 2007. 'Séminaire "Entrepôts de stockage, entrepôts et marchés. Pour une typologie des *horrea* dans l'Afrique du nord antique", 13 Octobre 2006, Maison Méditerranéenne des Sciences de l'Homme, Aix-en-Provence'. *Antiquités Africaines* 43:165–264.

Virlouvet, Catherine. 2011. 'Les entrepôts dans le monde romain antique, formes et fonctions. Premières pistes pour un essai de typologie'. In *Horrea d'Hispanie et de la Méditerranée romaine*, edited by Javier Arce and Bertrand Goffaux, 7–22. Collection de la Casa de Velázquez 125. Madrid: Casa de Velázquez.

Virlouvet, Catherine. 2015. 'Les métiers du port : les *saccarii*, dockers du monde romain antique [Review of: MARTELLI ELENA, *SULLE SPALLE DEI SACCARII. LE RAPPRESENTAZIONI DI FACCHINI E IL TRASPORTO DI DERRATE NEL PORTO DI OSTIA IN EPOCA IMPERIALE* (BAR International Series 2467; Archaeopress, Oxford 2013). Pp. 140, Tav/Pls. XIII. ISBN 978 1 4073 1078 7.]'. *Journal of Roman Archaeology* 28: 673–83. doi:10.1017/S1047759415002925.

Vos, Wouter K. 2002. *De inheems-Romeinse huisplattegronden van De Horden te Wijk bij Duurstede*. ROB rapportages archeologische monumentenzorg 96. Amersfoort: Rijksdienst voor het Oudheidkundig Bodemonderzoek.

Vos, Wouter K. 2009. *Bataafs platteland. Het Romeinse nederzettingslandschap in het Nederlandse Kromme-Rijngebied*. Nederlanse Archeologische Rapporten 35. Amersfoort / Amsterdam: Rijksdienst voor het Cultureel Erfgoed / Vrije Universiteit Amsterdam <http://hdl.handle.net/1871/13283>.

Vos, Wouter K. 2015. 'Identifying Discharged Soldiers in the Batavian Countryside'. In *Limes XXII. Proceedings of the 22nd International Congress of Roman frontier studies, Ruse, Bulgaria, September 2012*, edited by Lyudmil Ferdinandov Vagalinski and Nicolay Sharankov, 451–57. Izvestiâ na Arheologičeskiâ institut 42. Sofia: National Archaeological Institute with Museum of the Bulgarian Academy of Sciences.

Vossen, Ivo. 2007. 'Landschap, bewoning en landgebruik in de Romeinse tijd rondom Tiel'. In *Een Bataafse gemeenschap in de wereld van het Romeinse rijk. Opgravingen te Tiel-Passewaaij*, edited by Nico Roymans, Ton Derks, and Stijn Heeren, 33–44. Utrecht: Matrijs.

Vossen, Ivo, and Maaike Groot. 2009. 'Barley and Horses: Surplus and Demand in the *Civitas Batavorum*'. In *TRAC 2008. Proceedings of the eighteenth annual Theoretical Roman Archaeology Conference which took place at the University of Amsterdam, 4–6 March 2008*, edited by Mark Driessen, Stijn Heeren, Joep Hendricks, Fleur Kemmer, and Dirk Visser, 85–100. Oxford: Oxbow books. doi:10.16995/TRAC2008_85_100.

Wabont, Monique, Franck Abert, and Didier Vermeersch. 2006. *Le Val-d'Oise*. Carte archéologique de la Gaule 95. Paris: Académie des Inscriptions et Belles-Lettres.

Wagner-Roser, Silvia. 1987. 'Ein römischer Steckkalender aus Bad Rappenau, Kreis Heilbronn'. *Fundberichte aus Baden-Württemberg* 12: 431–38. doi:10.11588/fbbw.1987.0.39518.

Wamser, Gertrud. 1977. 'Ein römischer Gutshof bei Bad Rappenau, Kreis Heilbronn'. *Denkmalpflege in Baden-Württemberg – Nachrichtenblatt der Landesdenkmalpflege* 2 (2): 37–39. doi:10.11588/nbdpfbw.1973.2.14807.

Wesselingh, Dieke Antonette. 2000. *Native Neighbours. Local Settlement System and Social Structure in the Roman Period at Oss (The Netherlands)*. Analecta Praehistorica Leidensia 32. Leiden: University of Leiden. <https://hdl.handle.net/1887/33738>.

Wiethold, Julian. 2000. 'Kontinuität und Wandel in der landwirtschaftlichen Produktion und Nahrungsmittelversorgung zwischen Spätlatènezeit und gallo-römischer Epoche. Archäobotanische Analysen in der römischen Grossvillenanlage von Borg, Kreis Merzig-Wadern'. In *Kelten, Germanen, Römer im Mittelgebirgsraum zwischen Luxemburg und Thüringen. Akten des Internationalen Kolloquiums zum DFG-Schwerpunktprogramm 'Romanisierung' in Trier vom 28. bis 30. September 1998*, edited by Alfred Haffner and Siegmar von Schnurbein, 147–60. Kolloquien zur Vor- und Frühgeschichte 5. Bonn, Allemagne: R. Habelt.

Vos, Wouter K. 2003. 'How to trace the "Romanisation" of central Gaul by archaeobotanical analysis? – Some considerations on new archaeobotanical results from France Centre-Est'. In *Actualité de la recherche en histoire et archéologie agraires. Actes du colloque AGER V, 19–20 septembre 2000*, edited by François Favory and Anne

Vignot, 269–82. Annales littéraires de l'Université de Besançon / Environnement, sociétés et archéologie 764. Besançon: Presses universitaires franc-comtoises.

Wightman, Edith Mary. 1985. *Gallia Belgica*. London: B.T. Batsford.

Willcox, George H. 1977. 'Exotic Plants from Roman Waterlogged Sites in London'. *Journal of Archaeological Science* 4 (3): 269–82. doi:10.1016/0305-4403(77)90094-2.

Willems, Willem J. H. 1986a. 'De Romeinse Villa te Voerendaal. Opgraving 1985'. *Archeologie in Limburg* 28: 143–50.

Willems, Willem J. H. 1986b. 'Romans and Batavians. A Regional Study in the Dutch Eastern River Area'. Academisch Proefschrift, Amsterdam: Universiteit van Amsterdam. <https://openaccess.leidenuniv.nl/handle/1887/11881>.

Willems, Willem J. H. 1987a. 'De grote villa van Voerendaal'. In *Langs de weg: de Romeinse weg van Boulogne-sur-Mer naar Keulen, verkeersader voor industrie en handel. Villa Rustica, het Romeinse boerenbedrijf in het Rijn/Maasgebied*, edited by Petrus Johannes Joseph Stuart and Marjorie E. Th. de Grooth, 46–50. Heerlen / Maastricht: Thermenmuseum / Bonnefantenmuseum.

Willems, Willem J. H. 1987b. 'Voerendaal. Romeinse Villa'. *Publications de la Société d'Histoire et d'Archéologie du Limbourg* 124: 404–17.

Willems, Willem J. H. 1988. 'Die grosse villa rustica von Voerendaal'. In *Villa Rustica: Römische Gutshöfe im Rhein-Maas-Gebiet*, edited by Marjorie de Grooth, 8–13. Freiburg im Breisgau: Schillinger.

Willems, Willem J. H. 1992. 'Die kaiserzeitliche *villa* von Voerendaal = La *villa* romaine à Voerendaal = De Romeinse *villa* van Voerendaal'. In *Spurensicherung: archäologische Denkmalpflege in der Euregio Maas-Rhein = Relevés d'empreintes : la protection des vestiges archéologiques dans l'Eurégio Meuse-Rhin = Speurwerk : archeologische monumentenzorg in de Euregio Maas-Rijn*, edited by Gerhard Bauchhenss and Hans-Hoyer von Prittwitz und Gaffron, 526–34. Kunst und Altertum am Rhein 136. Mainz: Ph. von Zabern. <https://openaccess.leidenuniv.nl/handle/1887/10004>.

Willems, Willem J. H. 1995. 'Die Villa Rustica von Voerendaal (NL) und die ländliche Besiedlung'. In *Forschungen und Ergebnisse. Internationale Tagung über Römische Villen. Veszprém 16.-20. Mai 1994*, edited by Sylvia Palágyi, 116–23. Balácai Közlemények 3. Veszprém. <https://openaccess.leidenuniv.nl/handle/1887/10007>.

Willems, Willem J. H., and Harry van Enckevort, eds. 2009. Vlpia Noviomagvs, *Roman Nijmegen. The Batavian capital at the imperial frontier*. Journal of Roman Archaeology. Supplementary Series 73. Portsmouth (R.I.): Journal of Roman archaeology.

Willems, Willem J. H., and Laura I. Kooistra. 1987. 'De Romeinse Villa te Voerendaal. Opgraving 1986'. *Archeologie in Limburg* 32: 29–38.

Willems, Willem J. H., and Laura I. Kooistra. 1988. 'De Romeinse Villa te Voerendaal. Opgraving 1987'. *Archeologie in Limburg* 37: 137–47.

Wulfmeier, Johann-Christof, and Hans-Heinz Hartmann. 2009. 'Reichlich Speicherplatz. Ein horreum von Bad Rappenau, Kreis Heilbronn'. In *Landesarchäologie. Fest-*

schrift für Dieter Planck zum 65. Geburtstag, edited by Jörg Biel, Jörg Heiligmann, and Dirk Krausse, 341–78. Forschungen und Berichte zur Vor- und Frühgeschichte in Baden-Württemberg. Stuttgart: K. Theiss.

Zech-Matterne, Véronique, Ginette Auxiette, and François Malrain. 2013. 'Essai d'approche des systèmes agricoles laténiens en France septentrionale. Données carpologiques, archéozoologiques et archéologiques'. In *L'âge du Fer en Europe. Mélanges offerts à Olivier Buchsenschutz*, edited by Sophie Krausz, Anne Colin, Katherine Gruel, Ian B. Ralston, and Thierry Dechezleprêtre, 397–404. Mémoires – Ausonius 32. Bordeaux: Ausonius.

Zech-Matterne, Véronique, François Malrain, Stanislas Bossard, Thierry Lorho, Marie Derreumaux, Elsa Neveu, and Françoise Toulemonde. Forthcoming. 'Crop Productions, Formes and Sustainability of the Iron Age Farmsteads in Northern France: Differences and Convergences'. In *AGRIculture & LANDuse. Aspects of Acquisition and Economy of Rural Ressources in Iron Age Central Europe. Proceedings of the Annual Conference of the Iron Age Study Group (18–21 Septembre 2016, Münster)*.

Index of Ancient Sources and Inscriptions

Literary works are abbreviated according to the *Oxford Latin Dictionary* (online: <http://www.oxfordscholarlyeditions.com/page/abbreviations>) and the *Liddell-Scott Greek-English Lexicon* (online: <http://stephanus.tlg.uci.edu/lsj/01-authors_and_works.html>), supplemented by the *Index* of the *Thesaurus Linguae Latinae* where necessary. For Ausonius, use was made of the abbreviations in the ed. Green (1991).

Index of Place Names

www.ingramcontent.com/pod-product-compliance
Ingram Content Group UK Ltd.
Pitfield, Milton Keynes, MK11 3LW, UK
UKHW022209051026
15624UKWH00001B/80

9 789004 389038